The age and construction of English homes

A guide to ageing the English housing stock

Simon Nicol, Chris Beer and Chris Scott

The research and writing for this publication has been funded by BRE Trust, the largest UK charity dedicated specifically to research and education in the built environment. BRE Trust uses the profits made by its trading companies to fund new research and education programmes that advance knowledge, innovation and communication for public benefit.

BRE Trust is a company limited by guarantee, registered in England and Wales (no. 3282856) and registered as a charity in England (no. 1092193) and in Scotland (no. SC039320).
Registered office: Bucknalls Lane, Garston, Watford, Herts WD25 9XX
Tel: +44 (0) 333 321 8811
Email: secretary@bretrust.co.uk
www.bretrust.org.uk

IHS (NYSE: IHS) is the leading source of information, insight and analytics in critical areas that shape today's business landscape. Businesses and governments in more than 165 countries around the globe rely on the comprehensive content, expert independent analysis and flexible delivery methods of IHS to make high-impact decisions and develop strategies with speed and confidence. IHS is the exclusive publisher of BRE publications.
IHS Global Ltd is a private limited company registered in England and Wales (no. 00788737).
Registered office:
Willoughby Road, Bracknell, Berkshire RG12 8FB.
www.ihs.com

BRE publications are available from www.brebookshop.com
or
IHS BRE Press
Willoughby Road
Bracknell
Berkshire RG12 8FB
Tel: +44 (0) 1344 328038
Fax: +44 (0) 1344 328005
Email: brepress@ihs.com

© IHS 2014. No part of this publication may be reproduced or transmitted, in any form or by any means, electronic, mechanical, photocopying, recording or otherwise, or be stored in any retrieval system of any nature, without prior written permission of IHS. Requests to copy any part of this publication should be made to:
The Publisher
IHS BRE Press
Garston
Watford
Herts WD25 9XX
Tel: +44 (0) 1923 664761
Email: brepress@ihs.com

Printed using FSC or PEFC material from sustainable forests.

FB 71
First published 2014
ISBN 978-1-84806-361-7

All URLs accessed August 2014. Any third-party URLs are given for information and reference purposes only and BRE Trust and IHS do not control or warrant the accuracy, relevance, availability, timeliness or completeness of the information contained on any third-party website. Inclusion of any third-party details or website is not intended to reflect their importance, nor is it intended to endorse any views expressed, products or services offered, nor the companies or organisations in question.

Any views expressed in this publication are not necessarily those of BRE Trust or IHS. BRE Trust and IHS have made every effort to ensure that the information and guidance in this publication were accurate when published, but can take no responsibility for the subsequent use of this information, nor for any errors or omissions it may contain. To the extent permitted by law, BRE Trust and IHS shall not be liable for any loss, damage or expense incurred by reliance on the information or any statement contained herein.

Index compiled by Cathryn Pritchard.

Contents

Preface	iv
1 Age and type	1
2 Size, tenure, construction type, condition and location	6
3 Construction materials, street patterns, plot sizes and local history	13
4 Illustrated guide to the different periods in housing	22
5 References and other resources	106
Appendix: Age of building elements	108
Glossary of architectural terms	126

Preface

There are some 22.4 million homes in England, housing 52 million people. They range from the smallest studio flat to the largest palace, and vary in age from medieval castles built of local stone to new homes designed to the most exacting sustainability standards. Numerous books have been written on English housing – particularly on buildings of architectural interest. This guide is unique in that it provides detailed information on typical designs and features of houses built at different periods, using statistics from the 2010 English Housing Survey.

Experienced surveyors and housing professionals will have developed an intuitive feel for when a house was constructed and the date of any modifications that have been undertaken – particularly in their own local area, where they are familiar with local materials and designs. As such, this book is intended for less experienced professionals, students of housing or individuals with a personal interest, to be taken onto the streets and used as a reference book.

The book has been structured into four main sections. The first section defines what is meant by age – setting out different age bands from pre-1850 to the present day, and what is meant by type – such as detached, terraced and so on. The second section considers the factors of size, tenure, construction type, condition and location in identifying when a house was built. The third section provides the reader with tips on how to estimate the age of a house by looking at different aspects such as construction materials and individual elements, street patterns and plot sizes, and by finding out about the history of the area. The final section comprises an illustrated guide to different periods in housing, from pre-1850 to the present day.

1 Age and type

This section defines what is meant by age and type. It sets out the different age bands from pre-1850 to the present day, and defines the different types of houses such as detached, semi-detached and terraced.

The age and construction of English homes

By age, we mean the number of years that have passed since the construction date of the original building. The construction date is defined here as the year when the building in which the dwelling sits was first constructed. The building may have been considerably modified since that time, possibly involving changes in use, subdivision, conversion of a house to a flat or extensions to the original structure. For the purposes of this guide, we divide ages into the following bands:

- pre-1850 (historic)
- 1850–1899 (Victorian)
- 1900–1918 (Edwardian)
- 1919–1944 (interwar)
- 1945–1964 (early post-war)
- 1965–1980 (later post-war)
- 1981–2002 (modern)
- 2003–2010 (new).

By type, we refer to the following typology:

- detached house: a house designed with more than one storey above ground level, which is structurally independent of any neighbouring buildings. This includes chalet bungalows
- semi-detached house: a house of more than one storey, which is designed as part of a structural pair
- terraced house: a house of more than one storey, which is built as part of a structural terrace of at least three vertically separated units
- bungalow: a house with just one floor of living accommodation; usually detached but can be in pairs or even terraces
- converted flat: a flat formed by subdividing a house or other building into more than one unit
- purpose-built low-rise flat: a flat in a purpose-built block of fewer than six storeys
- purpose-built high-rise flat: a flat in a purpose-built block of six or more storeys.

The three-bedroom semi-detached house built between 1919 and 1964 is England's most common house type

What started out as a semi-detached bungalow built in the 1930s now has the appearance of a modern two-storey house. The clue to its original age and type is in the surrounding homes

A simple typology can be constructed with an 8 × 7 matrix using these defined ages and types together with the average floor area of each type (see Figure 1 overleaf). As can be seen from Figure 1, English homes are much more likely to be houses (80%) than flats (20%). Some house types were built in very large numbers in particular periods. The most common remaining pre-1850 house type is the rural detached house. There are still some 2.4 million terraced houses dating from before 1919. The most common house type built between 1919 and 1964 was the family semi, with some 3.5 million units still remaining. Since 1981 the housing stock has largely polarised into building small flats and houses for singles and couples, and detached houses for families.

Bungalows became popular in the twentieth century at a time when land was more easily available and densities were low, but it is no longer cost-effective to build them. Many have been converted into larger properties because they often sit on good plots in favourable locations, so it is likely that the number of true one-storey bungalows in the housing stock is reducing.

Converted flats are mainly created by subdividing larger urban houses built before 1919, although they can be created from buildings constructed for other uses, such as warehouses, offices, farm buildings and shops.

A corner shop with accommodation, now converted into two flats

A 1960s office block that has been converted into flats for sale and rent

Figure 1: A typology of English housing
(* The sample size in the English Housing Survey[1] data is too small to produce a reliable estimate.)

1 Age and type

	1945–1964	1965–1980	1981–2002	2003–2010	All ages
Small terraced	920,000 (79)	1,049,000 (80)	691,000 (71)	236,000 (93)	**6,356,000 (83)**
Medium/large terraced	1,754,000 (89)	922,000 (87)	539,000 (76)	126,000 (90)	**5,860,000 (94)**
Semi-detached	492,000 (147)	804,000 (133)	1,210,000 (134)	256,000 (153)	**3,796,000 (149)**
Detached	594,000 (75)	698,000 (77)	389,000 (91)	60,000 (80)	**1,996,000 (78)**
Bungalow	22,000 (*)	5,000 (*)	2,000 (*)	* (*)	**948,000 (65)**
Converted flat	541,000 (57)	938,000 (56)	797,000 (50)	370,000 (58)	**3,039,000 (56)**
Purpose built low-rise	73,000 (52)	187,000 (58)	23,000 (*)	74,000 (62)	**391,000 (58)**
Total	**4,397,000 (87)**	**4,602,000 (84)**	**3,650,000 (89)**	**1,112,000 (92)**	**22,386,000 (92)**

2 Size, tenure, construction type, condition and location

There are many factors that can help to identify a home as dating from a particular time, including dwelling size, tenure, construction type, condition and location. Each one of these factors is taken in turn and discussed in this section.

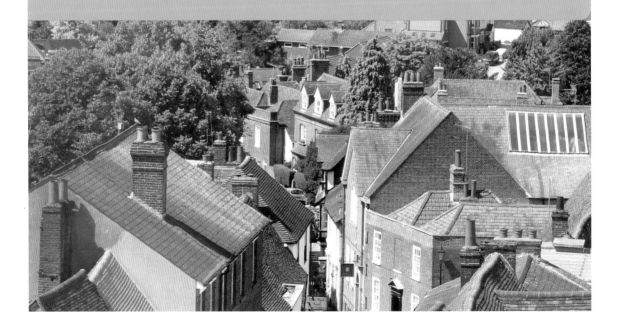

Dwelling size

The average useable floor area of the English housing stock is around 92 m², but this varies by house type and age. Detached houses built before World War I are generally larger than those built afterwards, but in recent years they have begun to grow again as plot sizes are maximised. It should be remembered that older housing that has been retained is generally of better quality than that which has been replaced. In addition, it is more likely that older housing will have been extended or modified from its original design than new housing.

Tenure

Large new detached house, making the most of a small plot by building up to the full extent of the boundary

Many house types are relatively tenure-specific (Table 1). In other words, some house types are most often built for a particular type of tenure, eg local authority or council houses, which are often readily identifiable by their utilitarian design. Council flats are often in 'deck-access blocks', which are a design almost universally provided for social housing. The great majority of council housing was built between 1919 and 1980. Some 1.8 million of these homes still exist. Since 1980 most new social housing has been built and managed by housing associations.

Privately rented houses, which used to consist mostly of poor-quality homes at the bottom end of the market, now include many more modern homes – either those specifically built to be let out at market rents (typically flats) or those provided for relocating business 'executives'.

Concrete-frame deck-access block, built for Coventry City Council in 1964, now transferred to a housing association

Table 1: Number of dwellings by age and tenure, 2010 (000s)[1]

Age	Owner-occupied	Privately rented	Local authority	Housing association	**All tenures**
Pre-1850	561	164	2	18	**744**
1850–1899	1,329	702	31	90	**2,153**
1900–1918	1,236	616	34	81	**1,967**
1919–1944	2,819	456	289	187	**3,751**
1945–1964	2,816	398	685	498	**4,397**
1965–1980	2,978	505	626	492	**4,602**
1981–2002	2,463	555	129	513	**3,650**
2003–2010	659	310	4	148	**1,122**
All ages	**14,860**	**3,706**	**1,801**	**2,018**	**22,386**

A Cotswold stone terraced house, c 1840

Oak-frame 17th century house, Hertfordshire

Construction type

The type and material of construction are good indicators of when a dwelling was built. Before the Industrial Revolution, building materials (which are heavy and difficult to transport) were sourced locally for all but the most expensive of homes. Often these local materials and building practices were very distinctive: Cotswold stone, hand-made brick, knapped flint, timber frames, tarred weatherboarding and so on.

Table 2 shows that the majority of homes built before 1850 had solid masonry walls (brick, stone, flint, cob), while the remainder had exposed hardwood timber frames. During Victorian times, mechanised production of bricks and joinery, combined with rail transportation, resulted in more uniform building materials, construction practices and designs. This era was dominated by solid masonry wall construction, typically of standard factory-made bricks. The cavity walls that exist from this period tend to be two leaves of stretcher bond with a narrow, untied gap between, mostly found in Lancashire. This arrangement was not for insulation purposes but to save on cost, as better-quality bricks could be used on the outside wall than on the inside.

True cavity walls were introduced after World War I and rapidly grew to become the most common type of wall construction. Modern cavity walls typically have a brick outer skin and a blockwork inner skin, although it is common for at least the first-storey outer skin to be rendered or clad blockwork in order to reduce cost and increase street appeal. Since the 1990s all new cavity walls contain in situ insulation. Of the remaining cavity wall stock, some 30% have been retrofitted with insulation.

The post-war periods saw increased experimentation in system- or factory-built homes, mainly for the local authority rented sector. This met a number of objectives, including: how to replace the homes lost

Table 2: Number of dwellings by age and construction type, 2010 (000s)[1]

Age	Solid masonry	Masonry cavity	Timber frame	Precast concrete	In situ concrete	Metal frame	All types
Pre-1850	691	–	54	–	–	–	**744**
1850–1899	1,970	175	8	–	–	–	**2,153**
1900–1918	1,617	347	4	–	–	–	**1,967**
1919–1944	1,646	2,057	17	6	21	2	**3,751**
1945–1964	303	3,694	33	138	182	47	**4,397**
1965–1980	47	3,984	117	72	355	27	**4,602**
1981–2002	38	3,243	249	10	62	47	**3,650**
2003–2010	13	854	113	10	73	58	**1,122**
All ages	**6,326**	**14,353**	**593**	**238**	**693**	**181**	**22,386**

'Lancashire cavity' walls on a 1909 terrace

Terrace of local authority houses with standard cavity walls, c 1950

during World War II; how to build new homes for returning servicemen; and how to use excess steel and concrete production, which was a legacy of the war effort. The various systems had names: the Cornish, Airey, BISF, Wates and so on. There is an exhaustive list of nearly 800 of these systems in a BRE publication, *Non-traditional houses*[2]. Around 1.5 million system-built homes were built before 1975 and around 400,000 of these are still in existence. Some of the more common types are illustrated below.

Such systems had gone out of fashion by the 1970s. They were only built with short life spans and some were already 35 years old, and certain components were starting to fail. A number of the systems built of concrete were designated as inherently defective under the Housing Defects Act 1984[3]. Many were demolished and replaced in the 1980s and 1990s, while others were comprehensively improved, sometimes involving replacement of the structure or overcladding with brickwork.

Concrete became the structural material of choice for large blocks of flats after World War II and some 700,000 flats in such blocks are now to be found. Steel is also increasingly used in large buildings containing flats and is, again, being used in modular housing designs.

Timber frames have again become popular, but instead of the historic style of a massive, usually exposed, hardwood frame with beams, they are factory formed in softwood and hidden behind a brick or blockwork skin, often with upper floors of panelling or tile hanging.

Distinctive Cornish Type 1 houses, built in the 1950s from a precast concrete factory kit frame and clad with concrete panels

Airey houses, built in the early 1950s, with a precast concrete frame and outer panelling

BISF houses, built with a steel frame and metal panelling in the early 1950s

Arcon steel bungalow, c 1948

New steel-frame modular home clad in PVC-U

In situ concrete-frame local authority block clad in brick, c 1964

New timber-frame house under construction

Condition

Building materials degrade, deteriorate or discolour over time, so condition can often be a good indication of date of construction. It is true that elements may be changed, often many times, during the lifetime of a building, so in order to properly identify the original construction date it is necessary to look at what appears to be the oldest remaining part.

Table 3 shows that older homes are more likely to be in disrepair than modern houses. A new home would not be expected to have any elements requiring repair, although it can clearly be seen from the table that even recently built homes have minor faults. The same pattern is followed through dwelling types and tenures, showing that the date of original construction is most significant when it comes to the state of repair.

Older homes are more likely to be in disrepair

Location

Pre-1919 housing is generally located in the centre of our cities, towns and villages (see Table 4 overleaf), or in isolated rural positions. Housing becomes increasingly modern as it is located away from the centre towards the suburbs. Nearly 70% of all homes built before 1850 are in rural locations.

Figure 2 overleaf shows the distribution of pre-1919 housing across the country, demonstrating its prevalence in rural, remote and inner city areas. Modern housing is particularly well represented in suburban and south-eastern areas. Figure 3 overleaf shows how housing of different ages is distributed across one English district. In this case, the oldest housing is concentrated in town centres and spread across rural areas.

Table 3: Average housing repair costs by age, 2010[1]

Age of construction	Mean basic repairs (£)
Pre-1850	4,208
1850–1899	2,549
1900–1918	2,543
1919–1944	1,847
1945–1964	1,333
1965–1980	757
1981–2002	469
2003–2010	123
All ages	**1,418**

Table 4: Number of dwellings by age and location, 2010 (000s)[1]

Age	Urban	Suburban	Village	Isolated rural	All locations
Pre-1850	135	96	322	192	**744**
1850–1899	953	785	281	134	**2,153**
1900–1918	971	775	158	63	**1,967**
1919–1944	658	2,670	332	61	**3,751**
1945–1964	511	3,168	634	84	**4,397**
1965–1980	690	3,144	714	54	**4,602**
1981–2002	532	2,407	645	66	**3,650**
2003–2010	274	636	192	19	**1,122**
All ages	4,724	13,710	3,278	673	22,386

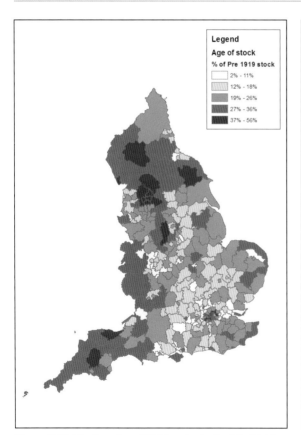

Figure 2: Distribution of pre-1919 housing by English district
Dwelling-age data provided under licence from Experian

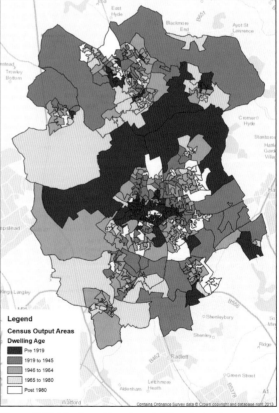

Figure 3: Distribution of housing of different ages in one English district
Contains Ordnance Survey data © Crown Copyright and database rights 2013
Dwelling-age data provided under licence from Experian

3 Construction materials, street patterns, plot sizes and local history

Estimating the age of buildings is not straightforward. It is therefore necessary to approach the task systematically and to make use of a wide range of information. It also makes sense to consider jointly the age of the building and its individual components. This section provides a set of tips to help to identify when a house was built.

The house on the left is a c 1875 conservation area terraced house; on the right is a new infill detached house in a 'pastiche' of the earlier style

These Cornish Mark II concrete-frame properties were both constructed in the 1950s but the one on the left was completely modernised, including the addition of a new brick outer skin, in the 1990s. Note the original roofs, chimneys and window openings, which are clues to the original date of construction

1. **Knowledge of the construction methods, typical materials, component designs and styles popular at different times is useful.** This is not foolproof, however. Some post-war homes were built to pre-war designs, for example. Some designs came in earlier or continued longer in some regions than others. Subsequent modernisations can make a home appear superficially as if it was built in a different era – usually a later era, although pastiches of earlier designs are not uncommon.
2. **The relationship between the age of the building as a whole and the ages of the individual components can be considered.** The components are very unlikely to be older than the building structure and may well be younger. In older buildings the individual components may have been replaced, but what are likely to have remained unchanged are the size and shape of the structure, the roof shape, the size and disposition of the windows, the wall materials and external detailing.
3. **Evidence of work having been done on the building may be helpful.** This might be indicated by a stylistic mismatch between one element and the building as a whole, or there might be physical indications of change, such as remnants of the replaced element still evident, a difference in mortar colour, evidence of a join between new and old, signs of age in some elements, degradation and moss/lichens on some elements but not on others.
4. **Comparison between a dwelling and its neighbours is very useful.** If the dwelling is on an estate it is likely to be a similar age to its neighbours. Some of the houses may have been modified in one way or another, but even on a fairly old estate there are likely to be a few properties that are in their original state.
5. **It can be useful to look at street patterns and plot sizes.** Victorian urban terraced houses are generally set out in straight parallel rows with narrow gardens. Interwar houses are found further from the town centre, often in estates with large gardens and spaces for garaging.

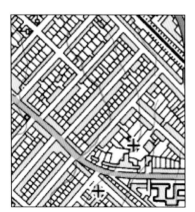

1850–1899 terraced housing with small yards

1919–1944 semi-detached housing with gardens

1981–2002 greenfield housing on former farmland

6. **Dated large-scale maps are very useful tools for ageing properties for those prepared to do a little research, as are aerial photographs.** For example, the series of London street maps clearly show its residential development through the years, while aerial photos show the characteristic linear street patterns of terraced houses and interwar suburbs. Other data sources exist that map out dwelling age profiles down to census output area.
7. **It is a good idea to find out about the history of the area.** Cathedral cities and market towns will still contain many medieval buildings in their core. Planned spa or resort towns (such as Bath, Leamington Spa and Brighton) will have publicised and recognisable housing from the Georgian/Regency periods. Letchworth and Hampstead have signature housing estates from the Garden Cities movement of the early twentieth century. Stevenage, Harlow, Telford and the other 'new towns' are heavily over-represented by housing from the 1965–1980 period. Milton Keynes is predominantly a post-1980 development. Local councils will often have information about conservation areas and locally listed buildings, while English Heritage has information about nationally listed buildings.
8. **Look for date stones and plaques.** Sometimes it is possible to find date stones or plaques on a building or on neighbouring buildings, stating the year in which the building was completed.
9. **It is useful to look at the layout of the house.** Pre-1919 terraced houses tend to be narrow, often with a smaller part built at the rear known as the 'back addition'. They also sometimes have a basement and an attic. Houses built after World War I are generally box-shaped and almost always of two storeys.
10. **A good way to age a house is by looking at the walls.** How thick are they? Are they solid or do they have a cavity? As a rule of thumb, homes built before 1919 will have solid walls, those built after 1945 are generally of cavity wall construction (see Figure 4 on page 18). Where the building is constructed from exposed brickwork, look at the bonding. Old English or Flemish bonding will generally confirm that the wall is solid. Stretcher bonding will usually confirm that it has a cavity.

Elegant Regency terrace, Belgravia, c 1825

Letchworth Garden City, c 1908

St Katharine Dock, London, 1980s–1990s

A pair of semi-detached houses, which by design could have been built at any time between 1900 and 1930. Note the original slate roof and sash windows (*top*). The actual date is confirmed as 1923 by the plaque (*bottom*)

New infill terrace built in the 1905 style of the area. Note the artificial slate roof and double-glazed sash windows and the cavity walls indicated by stretcher bond (*top*). However, it is the plaque that confirms the age of the building (*bottom*)

Blue plaques, which show when a famous or important person lived in a particular house, are often a good indicator of when it was built

An Edwardian end-terraced house. The street name (Ladysmith Road, a Boer War location) puts the date at 1902–1905

3 Construction materials, street patterns, plot sizes and local history

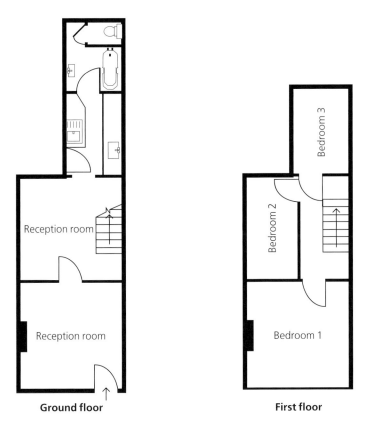

Typical floorplan for 1850–1899 terraced house with two-storey stepped back addition

Typical floorplan for 1919–1944 semi-detached house

18 The age and construction of English homes

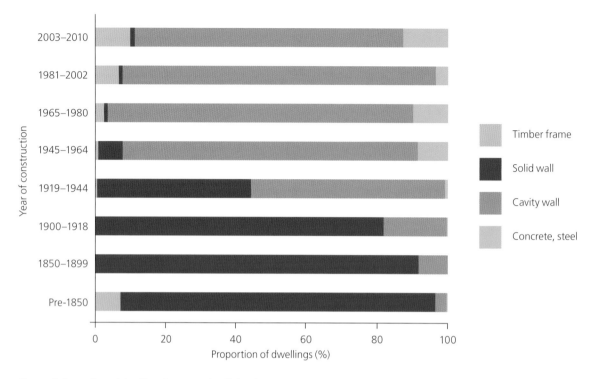

Figure 4: Proportion of dwellings by age and wall structure

This original art deco fireplace dates this house at around 1928

Kitchen from 1982

Bathroom from 1982

Brick bonding types

Bricks are arranged in different patterns so that they bond to provide suitable support. Most brick bonds are made up from a series of headers and stretchers in various configurations. The 'headers' are bricks laid to expose one end (or in a 9" wall, both ends). In contrast, a 'stretcher' brick is laid to expose one long face. The brick bonding types opposite represent the most commonly found in the English housing stock. In the English and Flemish bond examples, it should be noted that additional bricks (queen closers) are required to complete the bond. These 'quarter bricks' are required at either end of a run of brickwork and to maintain the bond at corners.

11. **It is also useful to look at the roof.** Roof designs, shapes, pitches and coverings have come in and out of fashion through the ages. (The eras of different roof types and materials are presented in the appendix.) Be aware that roof coverings have a limited lifespan and may have been replaced at least once during the life of the building. Thatched roofs will have been replaced many times since they were first installed. Figure 5 overleaf shows the proportion of the housing stock with original slate roofs by age.
12. **Looking at the windows and doors can help to age a house.** As with roofs, window designs, shapes, openings and materials have changed over the years: sash windows predominated between 1700 and 1919 (see Figure 6 on page 21); wooden bays with casements between 1919 and 1944; PVC-U double-glazed casements are the norm with new housing. See the appendix for further information.
13. **Information on the age of a house can also be gained from looking at the internal amenities, services and fittings.** If these are original they can offer good clues to the date of construction. If the dwelling still has the original heating system, this is a good clue to when it was constructed.
14. **Street names can help to date the houses on a particular road.** For example, all streets named after Boer War locations (eg Mafeking Road, Kimberly Road) will date from the early 1900s. Princess Diana Way could only date from the post-1980 period.
15. **Finally, it is a good idea to ask the occupier for information relating to the construction date of their home.** Most owners are surprisingly knowledgeable about when their homes were originally constructed, modernised and extended. This is particularly the case for long-term residents, some of whom may be the original occupier. Do not be afraid to ask. You will sometimes get the full history from an enthusiastic owner.

Solid 9" wall, old English bond, c 1875 (a row of headers followed by a row of stretchers)

Solid 9" wall, Flemish bond, c 1920 (rows of alternating headers and stretchers)

Cavity wall, stretcher bond, c 1970 (composed entirely of stretchers set in rows, offset by half a brick)

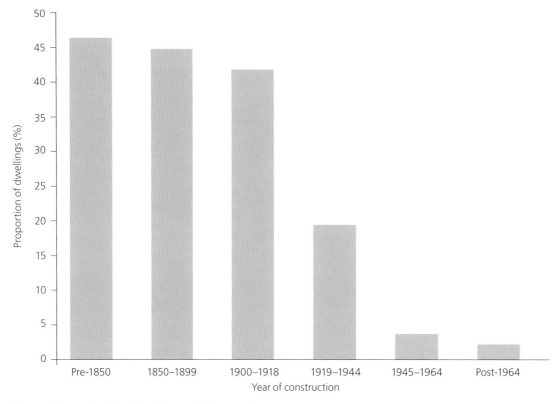

Figure 5: Proportion of dwellings by age with slate roofs

The house on the left retains the original slate roof tiles, while the house on the right has a replacement artificial slate roof

3 Construction materials, street patterns, plot sizes and local history

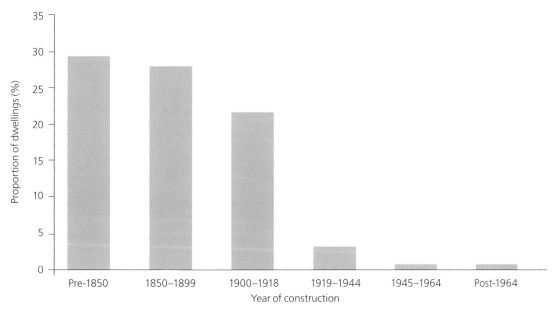

Figure 6: Proportion of dwellings by age with sash windows

Original sash window on pre-1850 house

Modern double-glazed sash window, c 1993

4 Illustrated guide to the different periods in housing

This section comprises an illustrated guide to different periods in housing, from pre-1850 to the present day. Each period is illustrated by a typical semi-detached house from the period showing many of its original features.

4 Illustrated guide to the different periods in housing

Pre-1850 (historic)

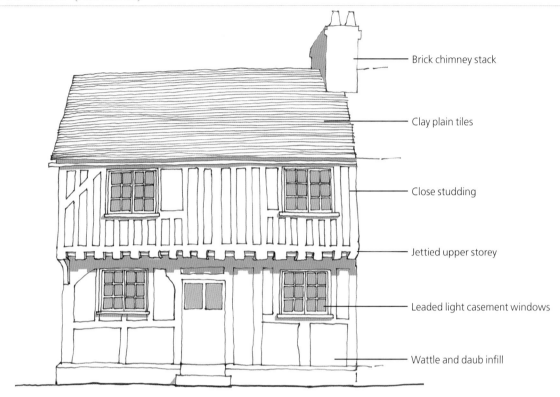

Typical historic semi-detached house

There are some 750,000 homes in England that date back to before 1850. These might be considered to be historic homes, and indeed around 250,000 are architecturally listed as such (the criteria for listing includes all dwellings built before 1700 that survive in anything like their original form, along with many of those built between 1700 and 1850 but with an element of selection involved). Even modest cottages that remain from this period will be at the upper end of the market, as generally only the better (rather than the typical) housing has been retained. Peasants' cottages, built of wood and mud in medieval times, have long gone, leaving only castles, substantial houses and attractive cottages remaining from this era, and perhaps leaving us with a rosy picture of how 'Olde Englande' used to look.

The historic period covers around 1,000 years of building, comprising several recognised architectural timescales that are grouped together in this publication, not because of their lack of interest or significance, but because of their limited numbers compared with the national stock of housing. There are other publications that focus exclusively on the style of, for example, Regency architecture. Historic houses fall broadly into the following periods:

- Saxon/Viking (450–1066)
- Norman/early English (1066–1485)
- Tudor/Elizabethan (1485–1603)
- Jacobean/Stuart (1603–1689)
- William and Mary/Queen Anne/early Georgian (1689–1760)
- Georgian (1760–1800)
- Regency (1800–1830)
- William IV/early Victorian (1830–1850)

24 The age and construction of English homes

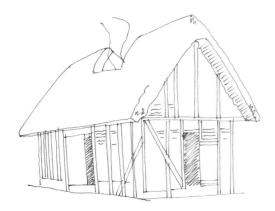

Saxon cottage, c 1000

Georgian townhouse, c 1775

Tudor house, c 1550

Regency terrace, c 1825

4 Illustrated guide to the different periods in housing

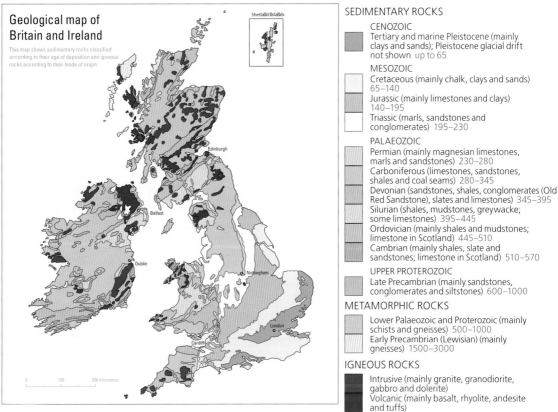

Figure 6: Simplified geological map of England and Wales
CP13/081 British Geological Survey © NERC. All rights reserved

Figures indicate age in millions of years

SEDIMENTARY ROCKS

CENOZOIC
- Tertiary and marine Pleistocene (mainly clays and sands); Pleistocene glacial drift not shown up to 65

MESOZOIC
- Cretaceous (mainly chalk, clays and sands) 65–140
- Jurassic (mainly limestones and clays) 140–195
- Triassic (marls, sandstones and conglomerates) 195–230

PALAEOZOIC
- Permian (mainly magnesian limestones, marls and sandstones) 230–280
- Carboniferous (limestones, sandstones, shales and coal seams) 280–345
- Devonian (sandstones, shales, conglomerates (Old Red Sandstone), slates and limestones) 345–395
- Silurian (shales, mudstones, greywacke; some limestones) 395–445
- Ordovician (mainly shales and mudstones; limestone in Scotland) 445–510
- Cambrian (mainly shales, slate and sandstones; limestone in Scotland) 510–570

UPPER PROTEROZOIC
- Late Precambrian (mainly sandstones, conglomerates and siltstones) 600–1000

METAMORPHIC ROCKS
- Lower Palaeozoic and Proterozoic (mainly schists and gneisses) 500–1000
- Early Precambrian (Lewisian) (mainly gneisses) 1500–3000

IGNEOUS ROCKS
- Intrusive (mainly granite, granodiorite, gabbro and dolerite)
- Volcanic (mainly basalt, rhyolite, andesite and tuffs)

Tudor, Jacobean and other pre-1700 buildings should be readily identifiable in most cases, but in urban locations earlier timber-frame buildings often lie behind later brick facades. The Great Fire of London of 1666 led to regulations that required all urban buildings built after this time to be constructed of masonry rather than wood.

Prior to the Industrial Revolution, houses used to be built primarily of local 'vernacular' building materials (see page 25), determined by the underlying geology (Figure 6), and often in very localised styles. Understanding how a home has been constructed and the materials used will help to date a building, as the majority of such homes will date from before 1850.

Houses built between 1700 and 1850 are characterised by distinctive classical proportions. Facades generally lack detailed ornamentation, often being unified by stucco or paint. Larger houses are often decorated with features derived from Greek or Roman architecture (the classical orders), such as columns, pilasters, pediments, cornices, rusticated lower walls and lintels, and occasionally wrought ironwork.

Windows and doors were usually spaced singly and were flat-headed or rounded, the windows invariably being sashes with smallish panes. Roofs were generally low-pitched, sometimes hipped, and on the main elevation were often hidden behind a parapet. Window and door reveals were sometimes rendered and painted.

Internally, large houses had servant quarters in a basement and/or in a small attic in a non-shared back addition. Surviving homes from this period are nearly all privately owned, of a larger than average size and typically terraced or detached houses.

Examples of vernacular housing

House built of local slate, Charnwood Forest, Leicestershire, 1889

Rural cottage with thatched roof, c 16th century

Random rubble and brick flank wall of pre-1850 cottage, west Berkshire

Early Georgian townhouse with Venetian-style windows, c 1760

Jurassic limestone house, Cotswolds, Oxfordshire, c 1860

Terraced cottages, Hertfordshire, 16th century

Examples of pre-1850 (historic) housing

Infill cottage with 2.5 m frontage, c 18th century

Timber-frame cottages with jetty, c 17th century

Timber-frame cottages, c 17th century

Brick cottages, c 18th century

Timber-frame thatched terraced cottages, c 16th century

Brick cottage, c 18th century

28 The age and construction of English homes

Examples of pre-1850 (historic) housing

End-terraced Georgian house, c 1800

Terraced cottages, c 1800

Regency terrace, c 1820. Note iron porch detail

Thatched cottages, c 16th century

Cottages with timber frame behind render, c 17th century

Timber-frame cottages with jettied first floor, c 17th century

4 Illustrated guide to the different periods in housing

Examples of pre-1850 (historic) housing

Timber-frame cottages, c 16th century

Terraced cottages, c 1800

Almshouses, 1816

Timber-frame and render cottages, c 17th century

Regency townhouse, c 1820

Timber-frame cottage with Georgian windows, c 1720

Examples of pre-1850 (historic) housing

Timber-frame cottages, c 18th century

Townhouses with pargeting detail, c 18th century

Georgian detached house with stucco finish, c 1750

Georgian townhouses, c 1800

Early Georgian townhouse, c 1720

Timber-frame cottage, c 17th century

4 Illustrated guide to the different periods in housing

Examples of pre-1850 (historic) housing

Timber-frame cottages, c 16th century

Timber-frame cottages, c 16th century

Rebuilt oak-frame cottage, originally c 16th century

Farmhouse with timber frame, c 17th century

Much improved and modified William IV house, 1831

Farmhouse, c 1840

Examples of pre-1850 (historic) housing

Georgian townhouse, c 1770

Early Victorian house with classical detailing, c 1840

Georgian house with Venetian-style windows, c 1760

Beach pebble and brick merchants' house, c 1800

Single-storey almshouses, c 17th century

Semi-detached thatched cottage, c 18th century

4 Illustrated guide to the different periods in housing

Examples of pre-1850 (historic) housing

Conversion of c 18th century barn into a detached house

Tudor mansion house, c 16th century

Regency townhouses converted into flats, c 1822

Georgian townhouses, c 1800

Timber-frame flats over shops, 1637

Mews houses at the rear of Regency terrace, c 1820

Examples of pre-1850 (historic) housing

Stucco-faced house, c 1830

Gothic revival mansion, c 1802, now converted into flats

Terrace of stucco-faced townhouse, c 1820

Flats and offices in early Victorian townhouses, c 1840

Stucco-faced houses, c 1830, now converted into large flats

Converted flats/offices in Georgian townhouses, c 1760

1850–1899 (Victorian)

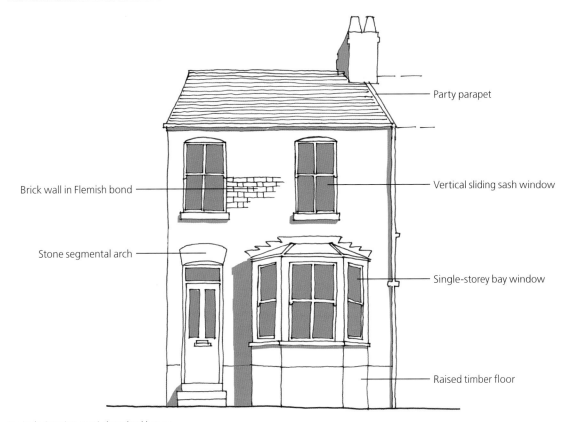

Typical Victorian semi-detached house

The Victorian era saw a rush to urbanisation throughout England, and almost every English town has an area of Victorian terraced housing surrounding the centre, where the population lived, largely segregated into classes according to their means and position in society.

The 2010 English Housing Survey[1] shows that there are still over one million terraced houses from this era, which were mainly originally constructed for landlords to rent out to the new urban workforce. These homes are still popular, but they are difficult to make energy-efficient and car parking is an issue.

Early Victorian homes often used narrow (or non-standard) bricks, and before 1870 lacked concrete foundations and damp-proof courses.

Mid- and late Victorian dwellings were characterised by elaborate decorative features. Facades often have more ornamentation than the equivalent pre-1850 homes, the detail being generally integral to the structural fabric, typically using contrasting ornamental brickwork or stonework at the edges of the facade or to pick out the windows and doors. Fired clay building components were used, in the form of both glazed and unglazed terracotta, as was reconstructed stone (eg in lintels). Decorative features often have a Gothic or medieval origin, such as indented cornices (the exception being the Queen Anne revival style where classical features were introduced as well).

Doors and windows were often arched, typically with a flat segmental arch. As well as being in bays, which were generally square or angular, windows were increasingly combined and, although still predominantly sashed, they had fewer panes. Roofs were steeper and generally visible, often having gables and attic dormers in the larger houses. Ridge tiles were often crenelated. Walls were still generally solid, although there was sometimes an outer front facade of better-quality brick or stone that was not fully bonded to the inner leaf, giving the appearance of a cavity wall. Internally, basements were less common, with servants' quarters being confined to larger, higher and often shared back additions.

The introduction of stricter building bye-laws gave more regularity, with walls in the main part being a minimum of 9" and floors being of the suspended timber type. Cities such as London (from 1707) and Bristol (from 1778) required fire-break parapets between adjoining terraced houses.

Urban Victorian terraced houses, c 1890. Car parking is always a problem!

The many streets of terraced houses were often named after patriotic events, locations and people, which can date them to a narrow period of time: Stanley and Livingstone after explorers; Victoria and Albert after the royal family; Balaclava and Omdurman after battles; Wellington, Beaconsfield and Gladstone after politicians.

During this period the first social housing created by charitable trusts began to appear, followed by that of pioneering local authorities such as the London County Council. For example, the Peabody Trust estates of London were almost invariably in the form of four- or five-storey blocks of flats with communal staircases and access galleries. Usually construction was of loadbearing brickwork, but less orthodox building elements, such as steel or cast iron joists or concrete columns, are also found.

Peabody Trust flats, Westminster

Towards the end of the nineteenth century the Arts and Crafts movement was born, which was essentially an artistic rebellion against industrialisation. Architects, interior designers and artists of the movement looked back to the craftsmanship of the pre-industrial age and brought this into house design, albeit initially at a very small-scale, upmarket and aesthetic level. These house designs had a major influence on the Garden Cities Movement that followed in the Edwardian era and the 'Tudorbethan' styling of the suburban housing of the early twentieth century.

An Arts and Crafts house from 1885 – a design hint of what was to come

Examples of 1850–1899 (Victorian) housing

Terraced houses, 1889

Terraced houses with basements, c 1875

Terraced houses, c 1865. Note original canopy detail

Terraced cottages with wooden bays, c 1885

Terraced houses with ornamental brickwork, c 1895

Stone terraced houses, c 1885

Examples of 1850–1899 (Victorian) housing

Brick and flint cottages, c 1890

Terraced houses with brick bays, c 1895

Stone cottages, c 1895

Terraced houses with parapet roofs, c 1885

Terraced houses with single-storey bays, c 1885

Stone terrace, c 1885

Examples of 1850–1899 (Victorian) housing

Slate cottages, c 1875

Terraced houses with single-storey bays, c 1885

Terraced houses, c 1898

Terraced houses, c 1885

Rural terraced cottages, c 1860

Semi-detached houses, c 1860

Examples of 1850–1899 (Victorian) housing

Terraced cottages, 1872. Note parapet wall to front and rear

Larger two-storey semi-detached houses, c 1895

Stone cottages, c 1880

Small terraced houses, c 1890

Victorian Gothic lodge house, c 1875

Cotswold stone detached house, c 1875

Examples of 1850–1899 (Victorian) housing

Gatehouse, c 1885. Note barge-boarding on gable

Late Victorian detached house, c 1890

Converted flats in large red brick house, c 1899

Detached village house built from random limestone rubble, c 1860

Semi-detached cottage, c 1890

Methodist hall, 1885, converted into a detached house

Examples of 1850–1899 (Victorian) housing

Early Victorian detached house, c 1850

Mansion house, c 1855

Much improved country cottage, c 1880

Terrace of tied bungalows, 1892

Semi-detached bungalows, c 1890

Converted flats in late Victorian townhouse, c 1895

Examples of 1850–1899 (Victorian) housing

Converted flats in early Victorian townhouse, c 1850

Converted flats in Victorian Gothic-style house, c 1880

Converted flats over shops, c 1898

Converted flats in 1890 house

Social housing block, 1888

Social housing block, 1897

Examples of 1850–1899 (Victorian) housing

Six storeys of flats and offices over shops, 1884

Mansion block, c 1895

Seven-storey mansion block, c 1895

Mansion block, c 1895

Converted warehouse development, 1895

Mansion block, c 1895

1900–1918 (Edwardian)

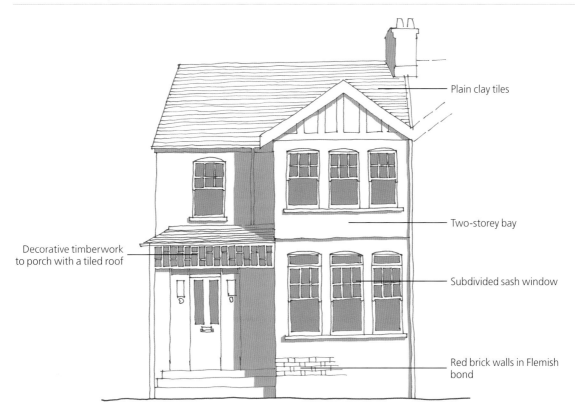

Typical Edwardian semi-detached house

Owner occupation for the middle classes was becoming the norm by the Edwardian era and Edwardian houses and villas were characterised by broader, more solid proportions to reflect people's aspirations. Larger houses were as likely to be built in semi-detached pairs as continuous terraces, hinting at what was to come during the interwar period. Facades tended to be more heavily ornamented than in the nineteenth century, but the ornament was often applied rather than integral with the structural fabric. Timberwork, tile hanging, the use of glazed tiles and terracotta were more frequently used, together with roughcast and art nouveau decoration.

Doors and windows were broader, with sashes being subdivided for ornamental reasons with leaded lights. Roofs tended to be slightly less steep but sometimes incorporated larger gables and decoration, such as timber finials. Clay finials and elaborately moulded ridge tiles were also used.

Wall tiling up to dado level was often used in a recess at the front entrance. This was sometimes extended through the entrance hall, which was often grander than the rest of the interior. Decorative, often black and white, floor tiles were used in the entrance hall. Stained glass was often used in the front door.

Internally, homes tended to be wider, and the rooms were larger but often fewer in number to reflect the fact that space for live-in servants was not generally required. With the increasing use of daily help and the consequent demise of the servants' quarters, back additions became smaller or less segregated, a greater emphasis being placed on the use of the back room and garden. Even modest houses designed for owner occupation had first-floor bathrooms. Smaller houses designed for tenants generally had outside WCs, but these were more accessible than their Victorian counterparts.

Nearly all houses had a damp-proof course combined with air bricks, which ventilated the suspended floors. Cavity walls started to become a feature of better-quality houses.

World War I brought a halt to housebuilding and very little was completed between 1914 and 1920, and what there was tended to follow later Edwardian styling.

Large Edwardian semi-detached houses, c 1910

Examples of 1900–1918 (Edwardian) housing

Three-bedroom terraced houses, c 1903

Three-bedroom houses, c 1902. Note roofs have been changed to concrete tiles from original slate

Three-bedroom single-bay end-terraced house, c 1905. Note rough-rendered upper floor

Terraced houses, 1909. Note older style but with narrow cavity walls

Terraced houses with cellars, c 1905

Terraced houses, c 1914

Examples of 1900–1918 (Edwardian) housing

Terraced houses, c 1907. Note larger window openings and recessed porches

Terraced houses, c 1910. Note wider frontages, square bays and large sash windows

Terraced houses, c 1905

Local authority homes, 1905

Garden City-style homes, c 1905

Garden City-style homes, c 1905

Examples of 1900–1918 (Edwardian) housing

Arts and Crafts-style terrace, c 1908

Arts and Crafts-style terrace, c 1910

Terrace, c 1905. Note steep roof pitch, typically with clay tile covering that was common in the period

Semi-detached houses, 1906

Semi-detached houses, c 1902

Semi-detached houses, c 1910

Examples of 1900–1918 (Edwardian) housing

Semi-detached houses, c 1910

Semi-detached houses, c 1908. Note square bays and half-timbered gables

Semi-detached houses, c 1908

Terrace of houses and converted flats, c 1910

Terraced houses, c 1910. Plain elevations, bathrooms and rendering are a sign of designs to come, but houses still have solid walls and back additions containing a scullery/WC

Semi-detached houses, c 1910, offering passage to the rear of the dwelling and side access creating a larger front room

Examples of 1900–1918 (Edwardian) housing

A pair of 'cottage-style' privately owned semi-detached houses, c 1910

Staggered terraced houses, c 1910. Clearly an intermediary stage between the Victorian terrace and the interwar semi

Larger-style semi-detached houses, c 1905

Larger-style semi-detached houses, c 1910

Larger-style semi-detached houses, c 1912. Note solid walls, lack of garage space and Garden City-style elevations

Staggered end-terraced house, 1910, with half-timbered bay, but still with solid walls and back addition with sash windows

Examples of 1900–1918 (Edwardian) housing

Larger-style semi-detached houses, c 1912. Note half-timbered effect, mixture of sash and casement windows, third floor

Larger-style privately built semi-detached houses, c 1913. Note solid wall construction, cat-slide roof, rough rendering to gable

Village detached house. Note in late Victorian style but built in 1906

Large detached house, c 1910. Note large sash windows

Large Garden City-style detached house, c 1913

Large detached house, c 1908

Examples of 1900–1918 (Edwardian) housing

Detached house, Hampstead Garden Suburb, c 1912

Bungalow, c 1905

'Pavilion'-style bungalow, c 1912

Tyneside flats, c 1910

Large house, now converted into flats, c 1910

Large house, now converted into flats, c 1910

Examples of 1900–1918 (Edwardian) housing

Purpose-built flats over shops, c 1912

Purpose-built flats over shops, c 1908

Privately owned flats, c 1903

Inner city social housing, c 1905

Six-storey privately owned block of flats, c 1905

Social housing, c 1905

1919–1944 (interwar)

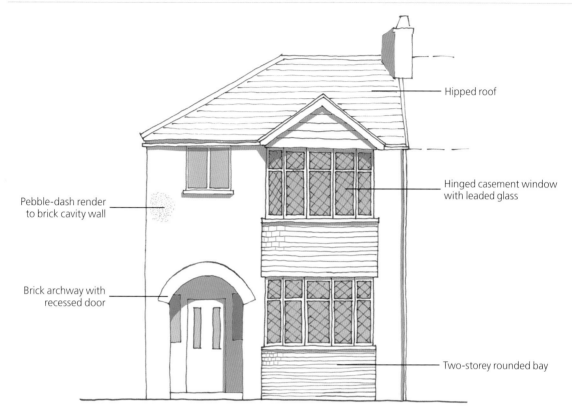

- Hipped roof
- Hinged casement window with leaded glass
- Pebble-dash render to brick cavity wall
- Brick archway with recessed door
- Two-storey rounded bay

Typical interwar semi-detached house

During this interwar period housebuilding boomed and styles changed dramatically. There were also strong stylistic differences between the homes designed for the growing numbers of owner occupiers and those for social housing tenants. Rent control, which was introduced during World War I, made building homes for private renting unattractive, and during this period private developers focused on building for sale to individual homeowners. The most common design was the semi-detached house (a uniquely British phenomenon); terraces were still used for smaller houses but these rarely comprised more than 4–6 units. This was the period of the large suburban estates, made accessible by good public transport ('Metroland' in London) and private car ownership. Often in 'Tudorbethan' styling and set in large plots, these reflected the aspirations of the growing middle classes. They were replicated on a smaller scale, without ornamentation and individualisation, on the new local authority estates that were built as 'homes fit for heroes' in the 1920s and 1930s.

To meet the demand for new housing, some local authorities turned to non-traditional (or system-built) techniques using steel frames, in situ concrete and precast concrete instead of bricks and mortar. Such houses were often rendered or brick-skinned and, as such, are difficult to distinguish from traditional masonry construction. This has been complicated by the fact that many of the remaining examples have had their original structures replaced by more traditional masonry and are now almost unrecognisable.

Art deco-style housing, c 1930

The ubiquitous 'semi' – 'Tudorbethan' semi-detached houses, c 1935

After 1930 much of the slum housing was cleared from our inner urban areas and replaced by local authority blocks of flats, typically walk-up blocks of four or five storeys. These flats were generally of loadbearing masonry construction, but some examples of steel-frame blocks exist in the major cities.

Traditionally constructed housing was typically built of masonry cavity walls. The homes designed for private occupation were often decorated with timbers set in render and they generally had a garage at the side. They had bay windows, often with leaded lights at the top and opening casements below. In the 1930s art deco styling became fashionable, typified by sweeping horizontally styled steel-frame windows and clean lines reflecting the aspirational 'Hollywood' look of the time.

Roofs of interwar houses tend to be hipped, rather than gable-ended. Roofs could be covered in slates or tiles, often reflecting regional and local preferences and availability. Internally, the doors were usually panelled and skirtings and architraves were moulded or bevelled.

The homes from this era have remained enduringly popular. They are generally well built, cavity walls and lofts can be made energy-efficient and their generous rooms offer flexible accommodation. The wide plots and large roof spaces offer plenty of opportunity for extension and adaption and there is usually enough room to park several cars. Their suburban locations still offer access to both town and country, schooling and transport links. The ubiquitous 'semi' is still our most common house type and is likely to be around for many years to come.

Examples of 1919–1944 (interwar) housing

Council housing, c 1935. Note external plumbing to bathroom, arch giving access to rear gardens

Council housing, c 1935. Note utilitarian concrete porch detail

Two-bedroom terraced house, c 1935

Small three-bedroom pebble-dash terrace, c 1928

Terraced houses with gable detail, c 1925

Double-fronted end-terraced council house, c 1930. Note original plain tile roof. Windows date from 1970s

Examples of 1919–1944 (interwar) housing

Two-bedroom semi-detached council houses, c 1925. Note original slate roof, replacement windows

Semi-detached houses, c 1938

Semi-detached houses, c 1938

Semi-detached houses, c 1938

Semi-detached houses, c 1920

Semi-detached houses, c 1930

Examples of 1919–1944 (interwar) housing

Semi-detached houses, c 1935

Semi-detached houses, c 1935

Small three-bedroom semi-detached houses, c 1935

Semi-detached houses, c 1935

Semi-detached houses, c 1920. Note older-style solid bays, slate roof and gable wall

Semi-detached houses, c 1930

Examples of 1919–1944 (interwar) housing

Semi-detached house in need of repair and maintenance, c 1930

Extended detached house, originally built c 1925

'Halls adjoining' semi-detached houses, c 1925

Semi-detached houses in 'Odeon' style, c 1935

Semi-detached houses, 1930

Semi-detached social housing, c 1928

Examples of 1919–1944 (interwar) housing

Local authority housing, c 1925

Semi-detached houses, c 1935

Detached house, c 1935

Detached house, c 1935. An unusual hybrid between traditional and art deco styling

Detached house with cavity walls and slate roof, c 1930

Detached house, c 1935

Examples of 1919–1944 (interwar) housing

Detached house, c 1938

Luxury detached house, c 1935. Completely renovated. Original age apparent from roof and chimney detail

Detached house, 1925

Detached house, 1935

Detached house in art deco style, c 1932

Large detached house, c 1928

Examples of 1919–1944 (interwar) housing

Semi-detached bungalow, c 1938

'Jerry built' bungalow, c 1925. Note original diamond-pattern roof tiles

Bungalow, c 1920. Note solid wall bays and half-timbered gables

Bungalow, c 1930. Note 'pyramid'-shaped hipped roof and wooden rounded bay

Bungalow, c 1935. Note original green tiled roof

Four-in-a-block maisonettes, c 1935

Examples of 1919–1944 (interwar) housing

Purpose-built flats, c 1930. Note horizontal-style windows, art deco detailing and flat roof

Deck-access walk-up social housing flats, c 1930

Walk-up block of flats, c 1928

Deck-access flats, c 1935

Mansion block, c 1938

Lutyens-designed deck-access flats, c 1930

1945–1964 (early post-war)

Typical early post-war semi-detached house

Another building boom followed World War II, particularly in the local authority sector. Around one million homes had been lost due to enemy action and many others had been damaged. At first progress was slow, clearance of bombed out sites being the priority, especially in cities like London, Coventry, Liverpool, Portsmouth, Southampton and Bristol. There was a shortage of bricks, mortar and construction skills, so many factories and workers that had been delivering steel, wood and concrete for the war effort were utilised to provide prefabricated homes in kit form for new local authority council estates. At first pre-war designs were used, but by the 1950s a more modern, utilitarian design became common.

The focus during this period was to provide everyone with a simple, basic home with all amenities and a garden in which to grow vegetables. Private-sector homes were built for owner occupation by working families and are thus, on average, smaller and more basic than the earlier homes built primarily for middle-class owners. Often large estates of private homes were built adjacent to, and are sometimes indistinguishable from, social housing estates.

Despite the proliferation of non-traditional house types during this period, most walls were usually of exposed cavity brickwork or a combination of this with various types of rendering. Towards the end of this period, timber boarding, tile hanging and concrete panels were used, particularly above ground-floor level. Window openings generally became wider than they were high; panes of glass tended to fill complete casements and large picture windows, of proportions hitherto unseen, became popular. Bay windows and ornamentation became less common. Homes of this era have thus become characterised by a rather plain and boxy appearance.

Slate roofs disappeared, to be largely replaced by concrete tiles. Some roofs, where flat or low-pitched, were covered in bituminous felt (usually with a green or grey grit surface). Roofs were usually gable-ended, but occasionally hipped.

During this period groups of flats, and mixed developments of houses and flats built in similar styles, became more common. These were usually walk-up blocks of 3–4 storeys. Towards the end of this era high-rise blocks of flats with concrete structures appeared with more regularity. Brutalist modern designs for 'cities in the sky' were imposed on social housing tenants in most of our large cities, eg Balfron Tower.

A 1950s detached house. Note the original steel windows and plain elevations

Balfron Tower, London, designed in the early 1960s by Erno Goldfinger and completed in 1967

Examples of 1945–1964 (early post-war) housing

Terraced council housing, c 1950

Terraced council housing, c 1955

Terraced council housing, c 1955

Terraced council housing, c 1960

Terraced council housing, c 1955

Terraced council housing, c 1953

Examples of 1945–1964 (early post-war) housing

Semi-detached council houses, c 1953

'Wates' concrete panel system council houses, c 1953

Semi-detached houses, c 1955

Semi-detached council houses, c 1955

Overclad semi-detached houses, c 1950

Semi-detached houses, c 1950

Examples of 1945–1964 (early post-war) housing

Semi-detached houses, c 1955

Semi-detached houses, c 1955

Semi-detached council houses, c 1952

Semi-detached houses, c 1950

Chalet bungalow, c 1963

Detached house, c 1955

Examples of 1945–1964 (early post-war) housing

Detached house, c 1950

Detached house, c 1960

Detached house, c 1955

Detached house, c 1955

Detached house, c 1958. Note replacement windows

Bungalow, c 1955. Note oriel window detail

Examples of 1945–1964 (early post-war) housing

Timber bungalow, c 1955

Terraced council bungalows, c 1955

Semi-detached bungalows, c 1955

Bungalow, c 1950

Flats designed to look like houses, c 1950

Flats designed to look like houses, c 1955

Examples of 1945–1964 (early post-war) housing

Local authority maisonettes, c 1955

Re-modelled maisonettes, originally built c 1960

Local authority flats, c 1955

Flats over shops, c 1955

Maisonettes, c 1960

Flats over shops, c 1955

4 Illustrated guide to the different periods in housing

Examples of 1945–1964 (early post-war) housing

Walk-up flats, c 1958

Walk-up flats, c 1955

Walk-up flats, c 1955

Walk-up flats, c 1960

Social housing block, c 1955

Mixed shops and flats, c 1960

Examples of 1945–1964 (early post-war) housing

Single concrete block of flats, c 1960

Re-modelled flats, c 1958

In situ concrete-frame tower block clad in brick, c 1964

Concrete tower block on estate, c 1958

Privately owned mansion block, c 1955

Brick-clad concrete block of deck-access maisonettes, c 1963

1965–1980 (later post-war)

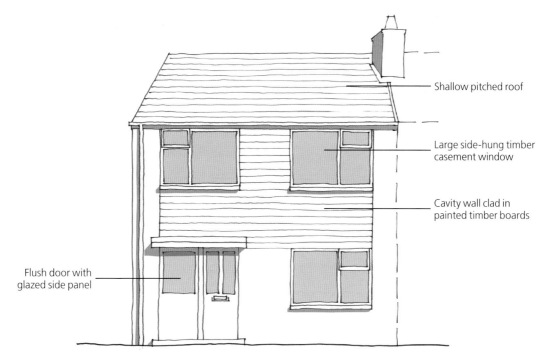

Typical later post-war semi-detached house

Design styles in this era are more varied than in previous periods. However, there was a proliferation of new suburban estates (both private and social) built on greenfield sites, which are of very distinctive design. They were often highly planned, with large areas of communal grassland, pedestrian walkways and garages tucked away out of sight in blocks. Houses were still boxy but usually made more cheery by the inclusion of painted panels, tile hanging and large windows. Roof elevations varied, using all forms of gabled, hipped and even mono-pitched slopes. Roofs covered with concrete tiles were of a lower pitch than in earlier periods.

During the 1960s low- and high-rise blocks of flats were built in a variety of types of construction – loadbearing brick, steel and concrete frames, and large concrete panel construction. Blocks of flats of this period were almost exclusively used to house local authority tenants.

This period saw the introduction of new materials on the outside of buildings: plastic gutters replacing asbestos cement, to be followed by plastic sidings. Windows were generally of the side-hung casement type, but occasionally pivoting windows, either vertical or horizontal, were used. Towards the end of the period plastics (PVC-U) became more widely used in window design, with casements predominating, although tilt-and-turn patterns following continental designs were also introduced. When thermal insulation requirements for external walls were made more demanding in 1975, windows began to decrease in size.

Internally, doors were normally of a flush pattern, and skirtings and architraves very plain with either a chamfered or a round profile. The ceilings were made of plasterboard, painted or with an Artex finish, sometimes with a coving at the junction of the ceiling and walls in the principal rooms.

Private-sector housing tends to evolve slowly in response to market trends. Such housing is a long-term investment, with buyers and financiers being understandably conservative. Many buyers prefer to invest in successful and robust existing housing rather than unproven new housing. On the other hand, local authority housing, particularly during the 1960s and 1970s, was an area of dynamic experimentation – in construction, style, layout and communal living. Typically, it was the middle-class planners and architects who lived in the proven private-sector areas who were designing these housing schemes for others to live in.

However, during the 1970s a number of factors came into play that made social housing less popular. There was a reaction against high-rise living, which meant single-family houses became more favoured in most new schemes. When flats were built, they tended to be on a more human scale; there was also a tendency to revert to smaller-scale street forms. An increasing disillusionment with more extreme forms of industrial building led to homes looking more like traditional housing, even when they had timber or concrete frames. The growing concern for energy conservation, and the implementation of energy efficiency requirements in the building regulations, led to a return to standard cavity wall construction with smaller windows. The rejection of large-scale demolition and renewal and the emphasis on rehabilitation, together with the growth of housing associations, led to mixed schemes combining newbuild with improvement, which reinforced the general move towards single-family homes. By the end of the 1970s, fewer greenfield sites were being allocated for large new estates, due to development control restrictions and public protest.

Low-rise system-built social housing flats, c 1968

Detached private house, c 1975

Examples of 1965–1980 (later post-war) housing

Small terraced houses with integral garages, c 1970

Terraced social housing, c 1978

Terraced housing, c 1970

Contemporary terraced housing, c 1975

Staggered terrace of social housing, c 1979

Terraced housing, c 1970

Examples of 1965–1980 (later post-war) housing

End-terraced house with integral garage, c 1978

Semi-detached house, c 1972

Extended detached house, c 1970

Architect-designed private house in contemporary style, c 1968. Note large windows and flat roofs

Wimpey-style semi-detached houses, mid-1960s. Note mixture of brick and render first-floor wall finishes

Wimpey-style semi-detached houses, mid-1960s. Note all rendered at first-floor level on the front face

Examples of 1965–1980 (later post-war) housing

Wimpey-style semi-detached houses, mid-1960s. Note tile-hung elevations

Wimpey-style semi-detached houses, mid-1960s. Note asymmetric pair of houses

Wimpey-style semi-detached houses, mid-1960s. Note crosswall style and narrower frontage with shared chimney

Wimpey-style semi-detached houses, mid-1960s. Note different-coloured facing brick to gable and front walls

Wimpey-style semi-detached houses, mid-1960s. Note all-brick finish

Wimpey-style semi-detached houses, mid-1960s. Note tiling over block at first-floor level

Examples of 1965–1980 (later post-war) housing

Plain detached house, c 1975

Detached houses, c 1975. Note large windows, low-pitched roofs and integral garages

'Georgian'-style detached house, popular in 1970s

Detached house, c 1970. Note replacement hardwood joinery

Extended and re-modelled house, c 1970

Staggered terrace with crosswalls and integral garages, c 1968

Examples of 1965–1980 (later post-war) housing

'Georgian'-style detached house, c 1975

Extended and re-modelled home, originally from c 1975

Large contemporary estate house, c 1975

Large detached house, c 1968. Note large windows

Georgian-style detached house, c 1975

Chalet-style house, c 1968

Examples of 1965–1980 (later post-war) housing

Four-bedroom detached house, c 1968

Georgian-style detached house using reclaimed bricks, c 1970

Semi-detached two-bedroom bungalows, c 1978. Note lack of chimney

Semi-detached bungalows, c 1965

Detached bungalow, c 1978

Semi-detached bungalows, c 1965

Examples of 1965–1980 (later post-war) housing

Detached bungalow, c 1965

Small block of flats, c 1975

Three-storey walk-up block of flats, c 1978

Two-storey corridor-access block of flats, c 1978

Walk-up block of flats, c 1965

'Brutal' concrete social housing flats, 1967

Examples of 1965–1980 (later post-war) housing

Brick-clad concrete block, c 1972

Social housing flats built in the mid-1970s incorporating section of the Byker Wall, Newcastle

Three 20-storey blocks. Originally built in 1965 and recently refurbished

City social housing complex with brick finish, c 1970

High-rise flats. Originally built in 1970 and refurbished in 2002

Sheltered housing block. Originally designed in 1975 and recently refurbished

1981–2002 (modern)

Typical modern semi-detached house

The lack of greenfield sites for building saw the number of housing completions steadily fall between 1981 and 2002, with some 3.65 million (mostly private-sector) constructed during this period. Local authorities sold off large swathes of their better housing stock over this period through the Right to Buy (RTB) scheme. These houses are easy to identify because their new owners often installed different-patterned windows and doors, and other signs of individuality, from their neighbours that are still in council ownership. (In more recent years, it is often the case that the homes still owned by the council have been comprehensively improved through Decent Homes schemes. In contrast, many of the RTB homes are now recognisable by their disrepair.) The large-scale transfer of whole council stocks to housing associations means that only 1.8 million homes remain in council ownership and virtually none have been built since 1980. There was significant building in the housing association sector, but this was mainly small developments on brownfield and infill sites. With land being in such scarce supply, housing densities increased and room sizes decreased. The remaining council estates have often been totally rehabilitated through Estates Action and other schemes and are now almost unrecognisable from their earlier form – pitched roofs built over concrete flat roofs, overcladding and decorative features will provide clues.

Between 1981 and 2002, the three million private-sector homes built were largely at either end of the affordability spectrum. Small, low-rise flats and starter homes proliferated to provide the first rung on the housing ladder for aspiring homeowners, while four-, five- or six-bedroom houses were built for the top end of the market. The quantity of existing Victorian terraced houses, interwar semis and post-war detached homes has been sufficient to meet the needs of private-sector families in the middle ground.

Both private and social housing built since 1980 generally has brick facades. Behind these facades, timber-frame construction is more prevalent, as the inner shell can be erected very quickly before being finished off by skilled bricklayers and fitters. Nevertheless, most owner occupiers still prefer to buy a home that looks traditionally built. Criticism of the monotonous external appearance of many developments led in the 1980s to an increase in the use of projecting porches, dormer and oriel windows, mock Tudor framing, false leaded lights and other features, coupled with a return to almost Victorian practices in the use of snapped headers in stretcher bond and patterned brickwork. There was, for a number of years, an increasing use of unsuitable tropical hardwood components for windows, doors, fascias and soffits, as well as PVC-U.

Fears over global warming, and the need to save energy and reduce carbon emissions, led to the introduction of new values for thermal insulation of external fabric, including windows in 1982, which had an effect on house designs. Signs of this, such as visible insulation blocks in roof spaces, can be seen in homes of this period. The government introduced the Standard Assessment Procedure (SAP)[4] in 1990 to facilitate the calculation of energy consumption in new buildings. As a consequence, double glazing became almost universal in new housing and windows tended to decrease still further in area. In order to accommodate sealed double-glazing units, glazing bars became thicker. Thermal insulation of solid and suspended ground floors has grown over the years to become the norm.

Larger-style 1990s terraced townhouse

4 Illustrated guide to the different periods in housing   87

Examples of 1981–2002 (modern) housing

Terraced houses, c 1982

Terraced houses with black wood cladding detail, 1981

Terraced houses, c 1982

Terraced social housing, c 1982

Contemporary terraced housing, c 1990

Terraced housing, c 1985

Examples of 1981–2002 (modern) housing

Semi-detached houses, c 1985

Semi-detached houses, c 1985

Semi-detached houses, c 1985. Note hardwood windows

Mixed small terraced houses and flats, c 1990

Four-in-a-block cluster home, c 1984

Semi-detached houses, c 1985

Examples of 1981–2002 (modern) housing

Mews-style houses in gated development, c 1995

Semi-detached houses, c 1985

Semi-detached houses, c 2000

Semi-detached houses, c 1989

Townhouses, c 2000

Townhouses with integral garages, c 2000

Examples of 1981–2002 (modern) housing

Townhouse with integral garage, c 1985

Detached house, c 1989

Detached house with integral garage, c 1990

Detached house in cottage style, c 1990

Detached house, c 1985

Detached house, c 1985

Examples of 1981–2002 (modern) housing

Detached house, c 1989

Experimental low-energy house, 1997

Detached house, c 1997

Detached house in Georgian style, c 1993

Detached house, c 1985

Detached house, c 1995

Examples of 1981–2002 (modern) housing

Bungalow, c 1985

Bungalow, c 1989

Bungalow, c 1989

Bungalow, c 1995

High-specification townhouses, c 1985

Infill development of small flats, c 1985

Examples of 1981–2002 (modern) housing

Infill development of small flats, c 1985

Townhouses and flats, c 1985

Walk-up block of flats, c 1985

Walk-up block of flats, c 1990

Walk-up block of flats, c 1990

Block of private flats, c 1995

Examples of 1981–2002 (modern) housing

Block of flats, c 2000

Small block of flats, c 1990

Infill urban flats, c 1995

Infill urban flats, c 2000

Pinnacle 1, Limehouse Basin, London, 1998

Block of flats, 1996

2003–2010 (new)

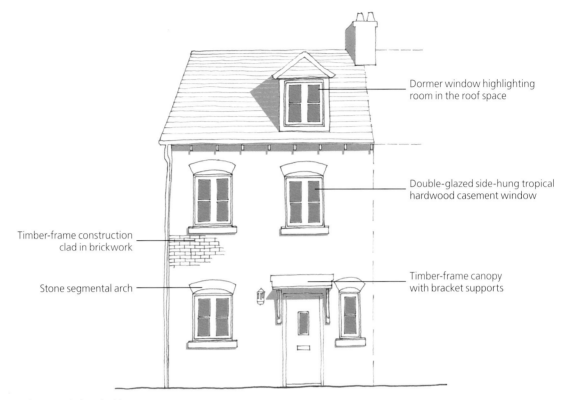

Typical new semi-detached house

In 2003 the housing market was booming, although the number of completions each year was still falling, largely because of planning restrictions and the lack of available land. The recession, which began in 2008 (and is only just coming to an end at the time of this publication), subsequently led to a stagnant housing market, with completions falling to around 100,000 units a year by 2010. Private-sector homes are still being built, particularly where there is a shortage of existing housing – mainly providing high-value, high-quality homes in good areas, largely on infill sites. The most expensive are often on gated developments, which are a growing phenomenon.

New social housing is provided by housing associations, usually to high quality standards but at increasingly high densities. Demolitions of existing homes are currently running at around 20,000 dwellings per year.

Ever more exacting energy efficiency standards for new housing have been introduced over the years, including the Code for Sustainable Homes[5]. All social housing must currently meet Level 3 of the Code for Sustainable Homes and there are prototype houses that can deliver Code Level 6, and are theoretically carbon-neutral. In order to encourage awareness of their energy performance, all new homes (and all existing ones sold) have been required since 2006 to have an energy performance certificate, which provides an energy rating based on SAP.

Stylistically, the housing developments being built today are much more individualistic than in the recent past, with no universal stereotypes such as those of the Victorian terrace, interwar semi or 1970s detached house. Planning control often dictates that new buildings on infill sites should blend in with the old, so

Exclusive new gated development of townhouses and flats

brand new energy-efficient homes will often take styling clues from the past. Looking from the street, the design of a traditional masonry home built by one builder might be indistinguishable from an off-the-peg timber-frame design built by another; it is only from the inside that the lightweight construction of the latter becomes apparent.

Bold futuristic developments are often found in showpiece brownfield developments, such as the dockland/canalside ventures of London, Manchester, Liverpool and Birmingham. While high-rise flats have fallen out of favour with the social housing sector, these expensive private-sector ventures often include industrial-style tower blocks, using concrete and steel clad in aluminium and glass in order to blend in with surrounding offices. Over 70,000 high-rise flats have been built since 2002, nearly all for the private sector and many as rental investments – a rate only equalled by the 1970s when the flats were for local authority tenants. The latest trend is for mixed-use 'signature' developments, with housing, offices, shopping and leisure facilities all in the same building complex. The Shard, completed in 2012, is the most extreme form of this development in the UK. It is 87 storeys high, with exclusive flats being located between the 53rd and 65th floors.

Recognising a new home should be straightforward. The home will probably still feel and smell new and fresh, be in good repair and have few obvious improvements or alterations. The planting on the development will not be mature. Densities will be high. Every inch of the available space will be used and homes will often appear too large for the plot, with a return to rooms built in the loft and multi-storey living. Single-storey bungalows have almost disappeared because of prohibitive land costs. New homes are generally built to sustainability standards (such as the Code for Sustainable Homes) and tropical hardwood, so common in joinery in the 1980s and 1990s, is now rarely used.

New low-energy house, Watford – the Sigma Home, built to Code Level 5

4 Illustrated guide to the different periods in housing

This new home looks to be of traditional construction, but is actually timber frame and clad in brickwork

The Shard, Southwark (a mixture of office, retail, leisure and housing facilities)

A group of new homes built at a much higher density in the back garden of a 1930s bungalow

Examples of 2003–2010 (new) housing

New two-bedroom starter home

New two-bedroom house. Note return of chimney to design

Small new three-bedroom semi-detached house

New terraced older-style townhouse. Note extra bedroom in loft space

Four-bedroom townhouse with traditional styling features

Three-bedroom social housing terrace, c 2005

Examples of 2003–2010 (new) housing

New infill terraced housing, styled to reflect existing Victorian houses

New two-bedroom terraced houses

Three-bedroom terraced houses in semi-rural setting

Three-storey traditionally built block of private flats

New end-terraced house with 'overflying' communal parking access

Plain new semi-detached houses

Examples of 2003–2010 (new) housing

Mixed townhouses and flats

Individual-style semi-detached houses

New semi-detached houses mimicking older style

Mixed townhouses and flats in traditional style

Low-energy traditionally built semi-detached houses

Mixed townhouses and flats in semi-rural location

Examples of 2003–2010 (new) housing

Three-storey townhouses with integral garages

Mixture of houses and flats in traditional style

Not a block of flats but a pair of large townhouses

Plain detached house squeezed into small plot

Larger-style semi-detached houses on greenfield site

Larger-style terraced houses on greenfield site

Examples of 2003–2010 (new) housing

New chalet in a street of bungalows

Another big house on a small greenfield plot, inspired by traditional designs but with incorrect proportions

New detached house with a mixture of new and traditional features. Note artificial slate roof and hardwood joinery

Mixed-size terraced housing in semi-rural location

Knock down a modest bungalow and build two enormous detached houses in its place!

Oversized detached house on small plot with random architectural features

Examples of 2003–2010 (new) housing

'Edwardian'-style detached houses on infill plot

Squeezing in new detached houses on greenfield site

Another large house on a small plot

Plain new house, but with lots of bathrooms

Executive new flats in grounds of stately home

High-density low-rise flats for first-time buyers

Examples of 2003–2010 (new) housing

'Chateau'-style block of flats on greenfield site

Large expensive flats in affluent area

New traditionally built flats for low-budget buyers

Cheerfully coloured new flats on urban infill site. Note use of green glass

Urban flats on brownfield site

Mixed flats and townhouses in semi-rural location

Examples of 2003–2010 (new) housing

Mixed-tenure flats on brownfield site

Mixed offices and flats in town centre

Large high-value flats

Mixed regeneration

Mixed development consisting of flats, offices, leisure and shopping facilities

High-profile high-value flats, 2004

5 References and other resources

References

1. Department for Communities and Local Government (DCLG). English housing survey: headline report 2010–11. London, DCLG, 2012.
2. Harrison H W, Mullin S, Reeves B and Stevens A. Non-traditional houses: identifying non-traditional houses in the UK 1918–75. BRE AP 294. Bracknell, IHS BRE Press, 2012.
3. Housing Defects Act 1984. London, The National Archives, 1984.
4. Department of Energy and Climate Change (DECC). SAP 2009: The government's standard assessment procedure for energy rating of dwellings. London, DECC, 2009.
5. Department for Communities and Local Government (DCLG). Code for sustainable homes: technical guide. London, DCLG, 2010.

Other resources

BRE. How to estimate the age of a dwelling. E-learning available at: www.bre.co.uk/eventdetails.jsp?id=4687.

English Heritage. The national heritage list for England. Available at: www.english-heritage.org.uk/professional/protection/process/national-heritage-list-for-england.

Prizeman P. Houses of Britain: the outside view. Wykey, Quiller Publishing Ltd, 2003.

Researching historic buildings in the British Isles. History of building regulations. Available at: www.buildinghistory.org/regulations.shtml.

Appendix: Age of building elements

The following tables show the main popularity periods for the particular form of construction. 'Hangs on' indicates that the method continued (or continues) to be used in new buildings after the 'finish' date, although not commonly. In some instances, where the popularity period is wide, a 'heyday' is given to the period when popularity was greatest.

It is also possible to date a property from the internal amenities and services, if these are still original. However, as kitchens and bathrooms tend to be replaced on average every 30 years or so it is not generally reliable to use this as the only indicator of the date of construction of the dwelling.

Beware, however, that obsolete or unfashionable materials and methods have continued to be used for replacement long after they stopped being used in new building, and that there is now something of a vogue for 'repro' components.

Chimney stacks

Description/appearance	Popularity period			Notes
	Start	Finish	Hangs on?	
Little guidance can be given on the dating of chimney stacks. Generally a stack will be of the same age as the building. Where it has been rebuilt, this will be seen (if at all) by signs of the work having been done, rather than by stylistic differences	Pre-1850	1970	–	Limited re-introduction after 1990 at higher end of market

Roof structure

Description/appearance	Popularity period			Notes
	Start	Finish	Hangs on?	
Pitched roofs (including mansard and chalet)				
With parapet; structure of rafters supported on purlins and roof plate				
Butterfly roof (valley gutter running front to back draining to the back, with a parapet in the front)	Pre-1850	1870	No	–
Conventional trussed roof behind a parapet	Pre-1850	1870	No	–
Gable/hipped; cut rafters and purlins on roof plate; earliest had deeper (5") rafters, later (after 1880) had 4" rafters; earliest had gables, later (during the 1930s) hipped became popular	Pre-1850	1950	No	–
Gable/hipped trussed roof; including 2–3 trusses and deeper purlins (generally in more substantial structures)	Pre-1850	Today	–	Heyday: up to 1965
Gable; factory-made trussed rafters (~15 rafters); earliest used bolted connections, later (after 1970) used toothed plates; earliest used 4" × 2" rafters, later (after 1980) used slimmer (~1.5") rafters	1965	Today	–	–
Steel; rolled steel joists used as purlins or for strengthening with wide spans	1910	Today	–	–
'System-built' roofs				
Composite wood/steel or fabricated steel rafters used in some building systems	1960	1970	No	Limited reintroduction after 1995
Concrete-frame members forming rafters and purlins used in some building systems	1960	1970	No	–

Roof structure continued

Description/appearance	Popularity period			Notes
	Start	Finish	Hangs on?	
Flat roofs				
Single family house				
Timber structure	1920	Today	–	–
Concrete; various construction methods (not accessible)	1930	1950	Yes	–
Flat blocks				
Timber or timber and steel	1890	1975	Yes	–
Concrete; various construction methods (not accessible)	1930	1975	Yes	–
Conservatories				
Wood; painted	1880	1950	Yes	–
Aluminium; single- or double-glazed; patented systems	1950	Today	–	–
Tropical hardwood; treated; single- or double-glazed (glass or polycarbonate)	1970	Today	–	–
PVC-U; single- or double-glazed	1975	Today	–	–
Accessories/fittings				
Ceilings				
Lath and plaster	Pre-1850	1940	No	–
Plasterboard	1930	Today	–	–
Roof lights				
Lantern roof lights	Pre-1850	1940	No	–
Lay lights (light at ceiling level with light at roof level above)	Pre-1850	1910	No	–
Cast iron roof lights	Pre-1850	1900	No	–
Wooden roof lights	1900	1960	No	–
Velux system; double-glazed	1960	Today	–	–
Dome; cast glass	1950	Today	–	–
Dome; acrylic	1960	Today	–	–
Dome; polycarbonate	1970	Today	–	–

Appendix

Thatched roof

Hipped roof

Mansard roof

Man-made slates (Eternit asbestos cement)

Natural slates

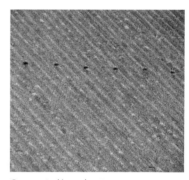
Corrugated iron sheet

Crenelated ridge tiles

Man-made slates (fibre cement)

Cedar shingles

Plain single-lap clay tiles

Concrete interlocking singe-lap tiles

Asphalt flat roof

Roof covering

Description/appearance	Popularity period			Notes
	Start	Finish	Hangs on?	
Pitched roofs				
Natural stone	Pre-1850	1935	Yes	–
Dressed slate	Pre-1850	1935	Yes	–
Wood shingles, usually cedar	1920	Today	–	–
Man-made slates				
Asbestos cement (Eternit); arranged as diamonds	1920	1940	No	–
Pressed steel; sand finished; comes in strips of four tiles	1975	Today	–	Used to replace slates during refurbishment because of similarity in weight
Resin; sometimes fixed by external clips to each slate, sometimes invisibly nailed	1985	Today	–	Often shows regular bowing, loss of surface coating, cracking around clips, edge discoloration, giving clues to age
Bitumen felt; comes in strips; 'slates' are thin, show black edges and have a granulated finish	1930	1980	Yes	–
Clay tiles				
Plain; pegged (oak pegs resting above battens or nails into battens)	Pre-1850	1890	No	–
Plain; nibbed; some earlier ones (before 1910) had curved edges	1870	Today	–	Heyday: up to 1960; paler-coloured tiles tend to be less durable
Clay; single-lap tiles				
Pan tiles (Roman tiles)	Pre-1850	Today	–	Heyday: up to 1960
Interlocking type	1900	Today	–	Heyday: up to 1960
Glazed (often bright green or blue)	1930	1940	No	–
Clay; double-lap interlocking tiles; fixed with clips at eaves and verges	1970	Today	–	–
Concrete tiles				Sand finish washes off with time – gives clue to age
Interlocking single-lap	1950	Today	–	–
Large plain 'Hardrow' single-lap (>A3)	1960	Today	–	Heyday: up to 1970
Plain double-lap	1950	Today	–	Heyday: up to 1970
Interlocking double-lap (with low pitch)	1965	Today	–	More angular profiles appear after 1975
Thatched; straw or reed	Pre-1850	Today	–	Ridges last 10–15 years; main thatch lasts 20–70 years; reed lasts longer than straw

Roof covering continued

Description/appearance	Popularity period			Notes
	Start	Finish	Hangs on?	
Pitched roofs continued				

Sheet
  Bitumen felt (occasionally used on flat blocks)
  Metal; corrugated iron (sinusoidal profile)
	1930
	1900
		1980
		1980
			No
			No
				–
				Heyday: up to 1940; used as replacement for thatch in rural areas

Metal; other profiles	1960	Today	–	–
Flat roofs				
Asphalt; with protective layer of chippings, tiles etc	1870	Today	–	–
Bitumen felt	1890	Today	–	–
Single-ply plastic membranes; smooth finish might be light-coloured, silver or white	1985	Today	–	–
Lead	Pre-1850	Today	–	Mainly used on dormers, bays and oriels
Sarking				
Torching; using cement/sand or lime/sand/hair	Pre-1850	1945	No	–
Timber featherboarding	Pre-1850	1940	No	–
Slaters felt; compressed felt saturated with bitumen (no warp and weft)	1920	1950	No	–
Untearable felt	1950	Today	–	–
Microperforated plastic sheet	1980	Today	–	–
'Tyvek'; the name is written all over it; non-woven compressed plastic fibre	1985	Today	–	–
Foam spray insulation	1980	Today	–	–
Accessories/fittings				
Ventilators in ridge, roof or eaves	1980	Today	–	–

Roof features and drainage

Description/appearance	Popularity period			Notes
	Start	Finish	Hangs on?	
Accessories/fittings				
Timber soffits/fascias	Pre-1850	1940	No	–
Asbestos cement soffits/fascias	1940	1970	No	–
Proprietary plastic soffits/fascias	1970	Today	–	–
Proprietary composite board soffits/fascias	1970	Today	–	–
Wooden decorative barge-boards	Pre-1850	1910	No	–
Valley gutters/flashings				
Little can be said that will assist dating	–	–	–	–
Gutters and downpipes				
Timber; often lead-lined guttering	Pre-1850	1880	No	–
Lead downpipes, guttering, hoppers, fittings	Pre-1850	1880	Yes	–
Cast iron downpipes and gutters	Pre-1850	Today	–	–
Concrete (Finlock system) built into the masonry structure and supporting the roof plate	1935	1980	No	–
Asbestos cement gutters and downpipes	1945	1970	No	–
Aluminium gutters and downpipes	1945	1965	Yes	–
Galvanised steel gutters and downpipes	1957	1970	No	–

Plastic downpipes and gutters
  Grey, half round
  Black/white/brown, half round
  Profiles other than half round
	
	1975
	1980
	1985
		
		Today
		Today
		Today
			
			–
			–
			–
				
				–
				–
				–

Stacks and wastes				
Location of soil pipe	–	–	–	Before 1965 soil pipes were fitted to the outside of the building. The Building Regulations 1965 (England and Wales) required that they be inside the dwelling, and the Building Regulations 1976 (England and Wales) allowed them on the outside again
Cast iron soil pipes	Pre-1850	1965	No	–
Plastic soil pipes	1975	Today	–	–

Roof features and drainage continued

Description/appearance	Popularity period			Notes
	Start	Finish	Hangs on?	
Party parapets				
Little can be said that will assist dating. Party parapets were required to give four hours of fire resistance for rows of four or more dwellings and two hours of resistance for rows of two or more, up until the introduction of building regulations around 1960	–	–	–	–

Wall structure

Description/appearance	Popularity period			Notes
	Start	Finish	Hangs on?	
Substantive part of structure				
Earth walls (eg cob, mud and stud, clay lump); tend to be found in localised areas	Pre-1850	1920	Yes	–
Stone				
Undressed (sometimes called 'rubble')	Pre-1850	Pre-1850	No	–
Dressed (called 'ashlar')	Pre-1850	Today	–	Look for date stone
Reconstituted stone (ground stone in concrete matrix, perhaps coloured)	1950	Today	–	–
Bricks and blocks				Look for date stone
Narrow bricks	Pre-1850	Pre-1850	No	–
Northern bricks (2 $^7/_8$", actual size)	Pre-1850	1970	No	–
Southern bricks (2 $^5/_8$", actual size) (became standard)	Pre-1850	Today	–	–
Moulded bricks (Victorian)	Pre-1850	1910	Yes	–
'Rubbed' bricks (soft, 'hand-rubbed' in construction of lintels)	Pre-1850	1910	No	–
Sand/lime (calcium silicate) bricks (white or lightly pigmented	1920	1995	Yes	Likely to be showing vertical cracking through bricks
Concrete blocks, 'self' finished (eg Forticrete) or rendered	1950	Today	–	–
Brick bonding and mortar joints				
Solid wall bonding (all types)	Pre-1850	1950	No	Look for date stone
Cavity wall bonding (normally all stretchers)	1935	Today	–	–
Narrow mortar joints	Pre-1850	1910	No	–

Wall structure continued

Description/appearance	Popularity period			Notes
	Start	Finish	Hangs on?	
Substantive part of structure continued				
In situ concrete; used in houses, low-rise and high-rise blocks	1920	1980	No	Most well known form is Wimpey no-fines, which was used 1945–1980
Precast concrete; used in houses, low-rise and high-rise blocks; system building using panels of various sizes, posts and beams	1960	1970	No	–
Framed structures				
Ancient timber frame; wattle and daub or herringbone brick infill	Pre-1850	Pre-1850	No	–

Timber frame; used in houses and low-rise blocks
  Early: timber-clad (narrow boarding)
  Later: brick-clad/rendered/resin-sprayed plywood panels
	
	1920
	1945
		
		1940
		Today
			
			No
			–
				
				–
				Heyday: 1965–1985; short gap then growth from 1990 and still growing today.

Steel frame; used in houses, low-rise and high-rise blocks; most easily identified in roof cavity; may be reclad in various finishes, perhaps several on one estate.	1920	1975	Yes	Heyday: 1955–1975
Accessories/fittings				

Lintels
  In situ concrete
 	
  Preformed concrete (square section)
  Preformed concrete (profiled section) (eg boot)
  Steel, straight
  Steel in shallow arch, used with soldier or snapped header bricks
	
	Pre-1850
	
	1920
	1950
	
	1970
	1970
		
		1950
		
		1950
		1970
		
		Today
		Today
			
			No
			
			Yes
			No
			
			–
			–
				
				Sometimes visible in 'less public' areas (eg garages)
				–
				Often visible on the facade
				
				Often visible on the facade
				Often visible on the facade

Coarsed rubble, undressed stone

Uncoarsed rubble, dressed stone

Ashlar

Brick

Small concrete panels

Large concrete panels

Tyrolean render

Pebble-dash render

Tile hanging

Unknapped flint

Softwood weatherboarding

Wattle and daub infill

Wall surface

Description/appearance	Popularity period			Notes
	Start	Finish	Hangs on?	
Stone finishes				
Knapped (dressed) flint	Pre-1850	1910	–	–
Natural (unknapped) flint	Pre-1850	Today	–	–
Brick and ornamentation				
Colour patterns: often bands or diamonds	Pre-1850	1900	Yes	Popular again since 1985
Gault white bricks	Pre-1850	1960	No	–
Fletton 　Textured surface 　Sand-faced	 1900 1930	 1960 Today	 Yes –	 – Heyday: 1940–1965
Concrete				
Blocks (eg Forticrete)	1950	Today	–	–
Panels (many finishes, eg exposed aggregate)	1960	1970	No	–
Pointing				
Labour-intensive styles (eg tuck pointed or galletted)	Pre-1850	1910	No	–
Narrow (< 5 mm)	Pre-1850	1910	No	–
Mortar				
Soft lime	Pre-1850	1930	Yes	–
Portland cement	1930	Today	–	–
Black ash (used in industrial areas and more generally)	1920	1950	No	If used in cavity walls will show cracking of bed joints on line of wall ties
Paints/surface treatments				
Cement-based (apparent from small shiny particles of mica)	1930	Today	–	–
Emulsion paint	1965	Today	–	–
Lime wash	Pre-1850	1960	Yes	Lasts around five years since it rubs off

Wall surface continued

Description/appearance	Popularity period			Notes
	Start	Finish	Hangs on?	
Renders				
Stucco (lime + ground brick); always painted, usually gloss oil paint	Pre-1850	1870	No	–
Smooth cement; sometimes painted	1850	Today	–	–
Pebble-dash	1920	1950	Yes	–
Tyrolean	1930	1940	No	–
Insulating renders (apparent from particles of insulant or from reinforcing fibres sometimes used on surface)	1980	Today	–	–
Applied surfaces				
Tile hanging				
Fancy edges to clay tiles	Pre-1850	1910	No	–
Plain clay tiles	1880	Today	–	–
Concrete tiles	1950	Today	–	Heyday: 1960–1980
Slate				
Natural	1850	1935	Yes	–
Artificial	1980	Today	–	–
Stone	Pre-1850	1935	No	–
Sidings (planking or shiplap)				
'Natural' wood lapped (with waney edges and bark) (eg elm)	Pre-1850	1940	No	–
'Finished' wood in vertical planking (eg cedar)	1920	1970	No	–
'Finished' wood in lapped horizontal planking	1920	1970	Yes	–
'Finished' wood in horizontal profiled interlocking planking	1950	Today	–	–
Plastic 'planking'	1985	Today	–	–
Sheeting				
Asbestos cement sheet	1920	1970	No	–
Plywood sheets (surface resin treated)	1950	1970	No	–
Profiled steel sheet	1945	1965	No	–
Profiled non-ferrous metal sheet	1950	1975	No	–
Plastic	1985	Today	–	–
Thin clay/brick slips	1970	1980	No	–
Thin artificial stone cladding	1985	Today	–	–

Softwood double-hung single-glazed sash window

PVC-U double-glazed sliding window

Triple-glazed sash window, c 2012

Tropical hardwood casements, c 1990

Stained glass insets

Side-hung double-glazed PVC-U, c 1998

Rolled steel (ungalvanised) window, c 1950

Softwood triple-glazed pivot window, c 2012

Projecting leaded single-glazed window

Side-hung leaded single-glazed window in single-storey box bay window

Side-hung single-glazed softwood casement in dormer

Side-hung single-glazed softwood casements

Windows

Description/appearance	Popularity period			Notes
	Start	Finish	Hangs on?	
Wood				
Indigenous hardwood (eg oak) frames and windows; unpainted	Pre-1850	1910	No	–
Softwood frames and windows; double-hung sash; painted				
Georgian: smallish panes (< A4); slender glazing bars	Pre-1850	Pre-1850	No	–
Victorian/Edwardian: one or two panes per sash; thicker glazing bars	1850	1915	No	–
Softwood frames and windows; casement; painted				
Interwar: panes (ie not 'picture' windows); heavyish glazing bars with intricate profiling	1920	1950	No	–
EJMA (English Joinery Manufacturers Association): panes and 'picture' windows; simple profiling with rounded arises on glazing bars and jambs	1950	1990	Yes	Heyday: 1950–1970; integral weather stripping from 1970
Tropical hardwood; single- or double-glazed	1970	1995	Yes	No longer considered sustainable
PVC-U				
All types				
Single-glazed	1970	1985	No	–
Double-glazed	1970	Today	–	–
Triple-glazed	2012	Today	–	Became more popular in 2012 although available since the 1990s

Windows continued

Description/appearance	Popularity period			Notes
	Start	Finish	Hangs on?	
Metal				
Cast iron	Pre-1850	1910	No	–
Aluminium				
'Mill finish' aluminium; sliding or casement; double-glazed; set in tropical hardwood sub-frames; no thermal barrier	1930	1970	No	Heyday: 1945–1965
White anodised or powder finish aluminium; sliding or casement; double-glazed; set in tropical hardwood sub-frames; no thermal barrier	1965	Today	–	Heyday: 1965–1980; thermal barrier introduced from 1980
White anodised or powder finish aluminium; sliding or casement; double-glazed; thermal barrier; set directly into masonry	1990	Today	–	–
Pressed steel over formed insulation in wooden sub-frames	1990	Today	–	–
Accessories/fittings				
Glazing (glass)				
Smallish panes (A4); curved ripples; bubbles	Pre-1850	Pre-1850	No	–
Larger panes; straightish ripples; bubbles	1850	1960	No	–
Float glass – few imperfections	1950	Today	–	–
Glazing (mounting)				
Leaded lights; rectangular panes in lead cames	Pre-1850	1910	No	–
Leaded lights; diamond panes in lead cames	1880	1940	Yes	–
'Leaded lights'; mock cames stuck on the glass	1960	Today	–	–
Stained glass: style of the period	1850	1940	Yes	–
Trickle ventilators (generally used)	1990	Today	–	Used in a few big windows (eg patio doors from 1975)
Ironmongery: of its period	–	–	–	May be newer than the windows but seldom older

Note: Window furniture may not be contemporary with main components. Generally the window is at least as old as the furniture.

Appendix

Oak door, c 1920

Softwood panelled door, c 1950

Indigenous hardwood door, c 1870

Painted softwood Victorian door

Painted softwood modern door

PVC-U triple-glazed door, c 2012

PVC-U front door and side panel

Natural oak front door, c 2010

Glazed softwood French doors with glazed side panels

Aluminium patio doors in tropical hardwood frame, c 1987

PVC-U double-glazed French doors, c 1995

Metal-frame patio door, c 1960

External doors

Description/appearance	Popularity period			Notes
	Start	Finish	Hangs on?	
Wood				
Indigenous hardwood (eg oak); framed and panelled	Pre-1850	1930	No	–
Softwood				
'Ledged, braced and boarded' or 'framed, ledged, braced and boarded' cottage doors	Pre-1850	1940	Yes	–
Painted; panelled: 'raised and fielded' panels (4, 6 or 8 panels within frame)	Pre-1850	1950	No	–
Painted; flush (plane) surface of plywood on frame, sometimes with small glazed opening	1940	1970	No	Heyday: 1950-1970
Painted; glazed; frame containing glazing or plywood panels, 2–3 in all	1950	1975	No	–
Tropical hardwood; treated or painted; panelled: 'raised and fielded' panels (4, 6 or 8 panels within frame) of various reproduction styles	1975	1995	Yes	Heyday: 1980s, but no longer considered sustainable
PVC-U				
All types				
Single-glazed	1970	1985	No	–
Double-glazed	1970	Today	–	–
Triple-glazed	2012	Today	–	Became more popular in 2012 although available since the 1990s
Metal				
Steel				
Rolled steel galvanised; casements; single-glazed; painted; set in wooden sub-frames normally	1950	1985	No	Paint often flaking off
Pressed steel over formed insulation; set in wooden sub-frames	1990	Today	–	–
Aluminium				
'Mill finish' aluminium; single- or double-glazed; set in tropical hardwood sub-frame; no thermal barrier	1960	1970	No	–
White anodised or powder finish aluminium; double-glazed; set in tropical hardwood sub-frame; no thermal barrier	1965	Today	–	Thermal barrier introduced from 1980

External doors continued

Description/appearance	Popularity period			Notes
	Start	Finish	Hangs on?	
Accessories/fittings				
Glazing (glass)				
Smallish panes (A4); curved ripples; bubbles	Pre-1850	Pre-1850	No	–
Larger panes; straightish ripples; bubbles	1850	1960	No	–
Float glass – few imperfections	1950	Today	–	–
Glazing (mounting)				
Leaded lights; rectangular panes in lead cames	Pre-1850	1910	No	–
Leaded lights; diamond panes in lead cames	1880	1940	Yes	–
'Leaded lights'; mock cames stuck on the glass	1960	Today	–	–
Stained glass: style of the period	1850	1940	Yes	–
Ironmongery: of its period				May be newer than the doors but seldom older

Note: Door furniture may not be contemporary with main components. Generally the door is at least as old as the furniture.

Damp-proof course

Description/appearance	Popularity period			Notes
	Start	Finish	Hangs on?	
Physical barrier				
Slate	1860	1910	Yes	Hangs on to 1940
Engineering-quality brick (called 'blue brick' or 'DPC brick') in two courses	1880	1940	Yes	–
Asphalt (thickness of normal mortar course)	1900	1920	No	–
Bitumen (thin course)	1900	1970	Yes	Normally 'weeps' on the sunny side of the building
Plastic	1965	Today	–	–
None				
None	Pre-1850	1880	No	–

Glossary of architectural terms

The following illustrations do not form a complete glossary of architectural terms, but are mentioned in this publication and are useful in helping to determine the age of a dwelling.

Elevation of historic (Tudor) house, c 1550

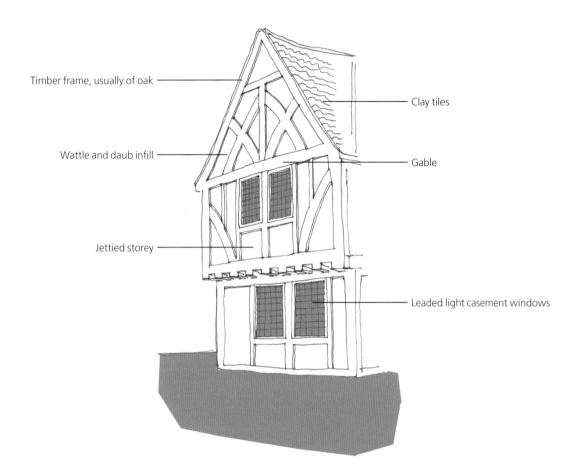

Clay tiles
Sometimes referred to as 'plain tiles'. Natural and durable tiles used in Britain since the thirteenth century.

Gable
The wall at the end of a ridged roof; generally triangular, sometimes semi-circular.

Jettied storey
A projecting upper storey.

Leaded light casement windows
Decorative windows made of small sections of glass supported in lead cames.

Timber frame, usually of oak
Sometimes referred to as 'post and beam', this form of house construction was popular in the Tudor period. Houses were created using heavy squared-off and carefully fitted and joined timbers with joints secured by large wooden pegs. It bears no relation to the modern method form of construction (MMC) of timber-frame dwellings, which uses standardised, prefabricated timber wall panels and floors.

Wattle and daub infill
An arrangement of small timbers (wattle) that form a matrix to support a daub. Daub is a sticky material made using different materials including soil, clay, limestone, chalk, stone, sand, straw and hair.

Exterior of Georgian townhouse, c 1790

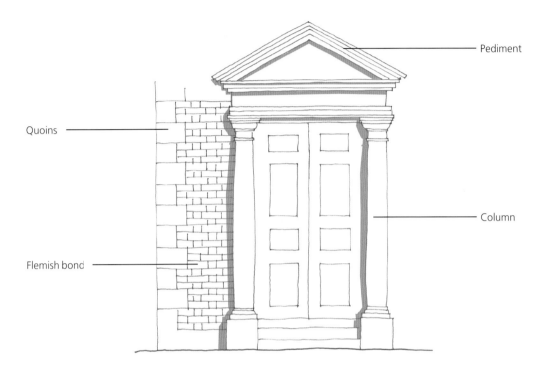

Column
An upright pillar, typically cylindrical, supporting an arch, entablature or other structure.

Flemish bond
A pattern of bricks in a wall in which each course consists of alternate headers and stretchers.

Pediment
A low pitched gable in classical architecture above a door, window or porch.

Quoins
The dressed stones or bricks at the angle of a building (pronounced 'coins').

Roof of London terraced house, c 1885

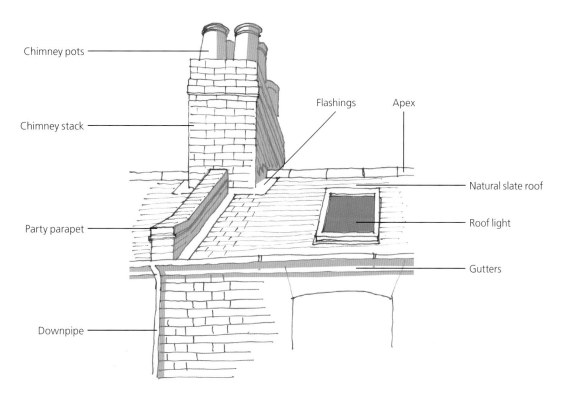

Apex
The peak, tip or highest point of the roof.

Chimney pots
A short pipe, normally of clay or metal. Chimney pots are placed on top of the chimneys to primarily improve the chimney's draft. A chimney with more than one pot on it indicates that there is more than one fireplace on different floors sharing the chimney.

Chimney stack
The part of the chimney often containing a number of flues, which projects above a roof.

Downpipe
A vertical pipe used for carrying rainwater away from the gutters to a drain or gully.

Flashings
Pieces of sheet metal used to protect and reinforce the joints and angles of a roof. Flashing is placed around discontinuities or objects that protrude from the roof of a building (such as pipes and chimneys or the edges of other roofs) to deflect water away from seams or joints and in valleys where the runoff is concentrated.

Gutters
Narrow troughs that sit above the roof fascia collecting rainwater and diverting it away from the structure.

Natural slate roof
Widely used as a roof tile in the late eighteenth century, the slate was mostly sourced from Welsh quarries. Today most of our slate is imported from Spain.

Party parapet
An extension of a party wall above the line of the roof surface. Primarily used to stop the spread of fire through the roof timbers.

Roof light
A flat or sloped window used for daylighting, built into a roof structure.

Internal door, c 1905

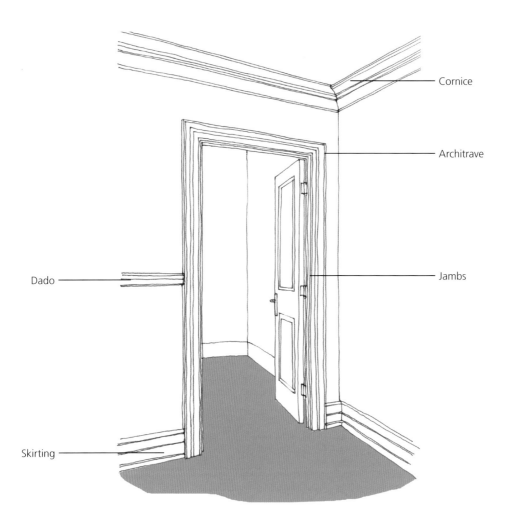

Architrave
A moulded enrichment to the jambs and head of a doorway or window opening.

Cornice
An ornamental moulding round the wall of a room just below the ceiling.

Dado
Short for 'dado rail', this is the lower part of the wall of a room, below about waist height, when decorated differently from the upper part.

Jambs
The sides of an archway, doorway, fireplace, window or other openings.

Skirting
A wooden board running along the base of an interior wall.

Glossary of architectural terms

Sash window in large block of flats, c 1925

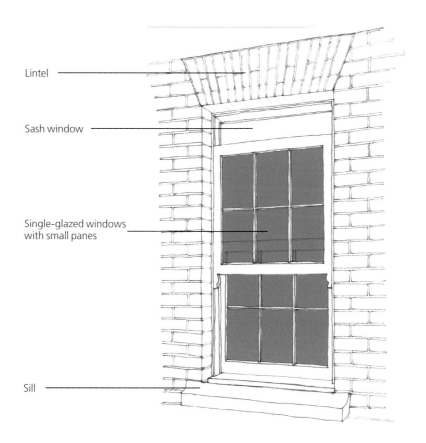

Lintel
The horizontal beam, brick or stone bridging an opening.

Sash window
A window with moveable panels or 'sashes' (often sliding vertically) that form a frame to hold the panes of glass.

Sill
A shelf or slab of stone, wood or metal at the foot of a window opening or doorway.

Single-glazed windows with small panes
The 'multi-lit' or 'lattice' windows with their small panes were the most popular type until the early twentieth century when industrial processes for glass making were perfected, allowing for larger panes of glass to be produced.

Lower part of cavity wall house, c 1965

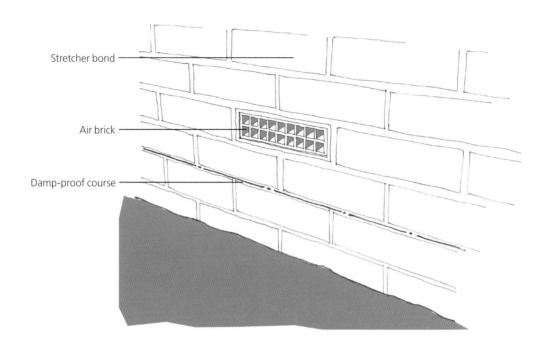

Air brick
A brick with holes allowing air to pass freely through a wall of a house, which aids ventilation and helps to prevent dampness. Often used below suspended timber floors.

Damp-proof course
A layer of waterproof material in the wall of a building near to the ground, to prevent rising damp. In pre-1945 houses this is usually two courses of bricks above ground level.

Stretcher bond
The pattern of bricks in the wall, where they are placed in a single unbonded row or in two rows with a cavity between them.

Glossary of architectural terms

Dormer window in roof of modern townhouse, c 1985

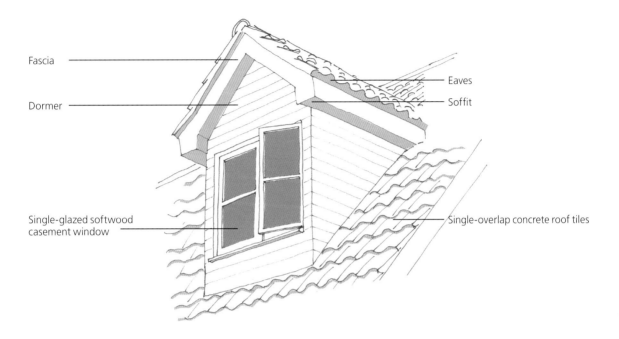

Dormer
A vertical window on the slope of a roof and having a roof of its own.

Eaves
The under part of a sloping roof overhanging a wall.

Fascia
A plain or moulded board covering either the rafter feet at the eaves or the plate of a projecting upper storey.

Single-glazed softwood casement window
A window with a hinged sash that swings in or out. It is the dominant type found in England.

Single-overlap concrete roof tiles
Single-lap tiles moulded to an S-shape, which gives the appearance of 'waves' and 'troughs' on the roof; often referred to as 'pantiles'.

Soffit
The underside of an architectural structure such as an arch, balcony or overhanging eaves.

Additional terms

Art deco
A decorative and architectural style of the period 1925–1940, characterised by geometric designs, bold colours and the use of plastic and glass.

Artex
A brand of coating for walls and ceilings that gives a textured finish.

Art nouveau
A style of decoration and architecture of the late nineteenth and early twentieth centuries, characterised particularly by the depiction of leaves and flowers in flowing, sinuous lines.

Ashlar
A thin dressed stone with straight edges that permit very thin mortar joints, used to face a wall.

Bonding
The regular arrangement of bricks in a pattern for strength or decoration.

Brutalism
An architectural style of the mid-twentieth century, characterised by massive or monolithic forms, usually of poured concrete and typically unrelieved by exterior decoration.

Came
A slender grooved lead bar used to hold together the panes in stained glass or latticework windows.

Chamfered
A flat surface made by cutting off the edge or corner of a block of wood or other material.

Close studding
A form of timberwork used in timber-frame buildings in which vertical timbers are set close together, dividing the wall into narrow panels.

Cob
Mud that is mixed with water, wheat straw and sometimes lime, which is then plastered over.

Corbel
A type of bracket made of stone, brick, wood or metal, which projects from the face of a wall to support a structure above it.

Crenelated ridge tile
A ridge tile that features a series of notches or indents, which gives the impression of a battlement.

Deck access
A block of flats that has a continuous inset balcony at each level onto which the front door of each flat on that level opens.

Estates Action
A government programme started in the mid-1980s to improve local authority housing estates.

Facade
An external face of a building. Usually, but not always, the principal or front face.

Finial
A sculptured ornament, often in the shape of a leaf or flower, at the top of a pinnacle, gable or similar structure.

Hipped roof
A roof that has sloping ends as well as sloping sides.

In situ concrete
A construction system that is produced 'on site' by casting it in its final location in the finished structure.

Lintel
A horizontal structural member, such as a wooden beam or stone, which spans an opening (as between the uprights of a door or window) and holds the weight of the structure above it.

Mansard roof
A four-sided roof having a double slope on all sides, with the lower slope much steeper than the upper.

Oriel window
A projecting bay window, often supported by corbels or brackets.

Pilaster
A shallow rectangular support that resembles a flat column that projects only slightly from the wall.

Precast concrete
A construction product produced 'off site' by casting concrete in a reusable mould, which is then cured in a controlled environment, transported to the construction site and lifted into place.

Right to Buy
A scheme that gives secure tenants of councils and some housing associations the legal right to buy, at a large discount, the home they are living in.

Rusticated
Cut or shaped large blocks of masonry that are separated by deep joints and decorated with a design to create a bold textured look.

Stucco
A form of cement render used to cover brick or stone.

'Tudorbethan'
A style of architecture that originated in the mid- to late nineteenth century, which amalgamates Tudor, Jacobean and Elizabethan styles.

Vernacular
An indigenous building style using local materials and traditional methods of construction and ornament.

Walk-up access
A form of access to flats within a multi-storey block whereby each flat is entered directly from a staircase landing.

Index

Page numbers in **bold** indicate figures and tables; those in *italics* indicate appendix and glossary items.

A

aerial photographs 15
age bands 2
ageing tips 14–21
air bricks 46, *132*
Airey system 9, **9**
almshouses 29, **32**
apex *129*
architect-designed houses 78
architraves 56, 75, *130*
Arcon construction 10
art deco-style housing **18**, 56, **56**, **61**, **62**, **64**, *134*
art nouveau decoration 45, *134*
artex 75, *134*
Arts and Crafts movement 36, **36**, 49
ashlar *115*, **117**, *134*
attics 15, 25, 36

B

Balfron Tower, London **66**
barge-boarding **41**
barn conversions **33**
basements 15, 25, 36, **37**, 47
bathrooms **18**, 46, 108
bay windows 19
 early post-war period (1945–1964) 65
 Edwardian period (1900–1918) **47**, **48**, **50**, **51**
 interwar period (1919–1944) 56, **59**, **63**
 Victorian period (1850–1899) 35–36, **35**, **37**, **38**, **39**
BISF system 9, **10**
blue plaques **16**
brick bonding 15, 19, *115*, *134*
 Flemish 15, 19, **19**, *128*
 Old English 15, 19, **19**
 stretcher 15, **16**, 19, **19**, *132*
bricks *115*, **117**, *118*

brownfield developments 96, **104**, **105**
brutalism 66, **66**, **83**, *134*
bungalows 2, **4–5**
 chalet **69**, **101**
 conversions **2**, 3
 early post-war period (1945–1964) **69**, **70**, **71**
 Edwardian period (1900–1918) **53**
 interwar period (1919–1944) **63**
 later post-war period (1965–1980) **82**, **83**
 modern period (1981–2002) **92**
 new homes (2003–2010) **101**
 'Pavilion'-style **53**
 timber **71**
 Victorian period (1850–1899) **42**

C

came *134*
car parking 35, **36**
carbon-neutral houses 95
casement windows 19, **120**, *121*, *122*, *133*
 early post-war period (1945–1964) 65
 Edwardian period (1900–1918) **52**
 interwar period (1919–1944) 56
 later post-war period (1965–1980) 75, **75**
 leaded light **120**, *122*, *127*
 metal-frame **65**, **66**, **120**, *122*
cat-slide roofs **52**
cathedral cities 15
cavity wall construction 8, **9**, 15, **19**, *115*, *132*
 early post-war period (1945–1964) 65
 Edwardian period (1900–1918) 46, **47**
 interwar period (1919–1944) 56, **61**
 later post-war period (1965–1980) 76

cavity wall construction (*continued*)
 numbers of dwellings **8**
 proportion by age **18**
ceilings 75, *110*
cellars *see* basements
chalet bungalows **69**, **101**
chalet-style houses **81**
chamfered *134*
charitable trusts 36
'chateau'-style blocks of flats **104**
chimney pots *129*
chimney stacks **98**, *109*, *129*
 shared **79**
classical features 25, **32**, 35
clay tiles **45**, **111**, *112*, *127*
cluster homes **88**
cob 8, *134*
Code for Sustainable Homes 95, 96
columns *128*
concrete canopies **57**, **65**
concrete construction 9, *116*, **117**
 numbers of dwellings **8**
 precast 55, **116**, **117**, *134*
 proportion by age **18**
 in situ **10**, 55, **74**, *116*, *134*
 see also system-built homes
concrete tiles **47**, 66, 75, **85**, **111**, *112*, *133*
condition 11, **11**
conservation areas 15
conservatories *110*
construction types 8–9
 numbers of dwellings **8**
 see also cavity wall construction; concrete construction; metal-frame construction; solid masonry wall construction; stone construction; timber-frame construction
converted buildings **33**, **41**, **44**
converted flats 2, 3, **3**, **4–5**
 Edwardian period (1900–1918) **50**, **53**
 historic period (pre-1850) **33**, **34**
 Victorian period (1850–1899) **41**, **42**, **43**, **44**
corbels *134*
cornices *130*
Cornish system 9, **9**, **14**

corridor-access blocks of flats **83**
Cotswold stone 8, **8**, **26**, **40**
cottage-style houses **51**, **90**
council housing 7, **7**, **10**
    Decent Homes schemes 85
    early post-war period (1945–1964) 65, 66, **66**, **67**, **68**, **69**, **71**, **72**
    Edwardian period (1900–1918) **48**
    Estates Action schemes 85
    interwar period (1919–1944) 55, 56, **57**, **58**, **61**
    later post-war period (1965–1980) 75, 76, **76**
    numbers of dwellings **7**
    Right to Buy (RTB) scheme 85
    Victorian period (1850–1899) 36
crenelated ridge tiles **111**, *134*

D

dados *130*
damp-proof courses 35, 46, *125*, *132*
date stones 15, **16**
Decent Homes schemes 85
deck-access blocks 7, **7**, **64**, **74**, *134*
decorative features
    early post-war period (1945–1964) 65
    Edwardian period (1900–1918) 45, **45**
    historic period (pre-1850) 25, **28**
    interwar period (1919–1944) 56
    modern period (1981–2002) 86
    Victorian period (1850–1899) 35, **37**
densities, housing 3, 85, 95, 96, **97**, **103**
detached houses 2, **4–5**
    early post-war period (1945–1964) **66**, **69**, **70**
    Edwardian period (1900–1918) **52**, **53**
    'Georgian'-style **80**, **81**, **82**, **91**
    historic period (pre-1850) **26**, **30**, **31**, **32**
    interwar period (1919–1944) **60**, **61**, **62**
    later post-war period (1965–1980) **76**, **78**, **80**, **81**, **82**
    modern period (1981–2002) **90**, **91**

detached houses (*continued*)
    new homes (2003–2010) **101**, **102**, **103**
    sizes 7, **7**
    Victorian period (1850–1899) **40**, **41**, **42**
doors 19
    external 25, 35, 45, **123**, *124–125*
    internal 56, 75, *130*
dormer windows 36, 86, **95**, **120**, *133*
double glazing
    doors **123**, *124*
    windows **16**, **21**, 86, **120**, *121*
downpipes *114*, *129*
drainage *114*
dwelling size 7

E

early post-war period (1945–1964) **5**, 65–66, **65**
    bungalows **69**, **70**, **71**
    construction types 8–9, **8**, **9**, **10**
    detached houses **66**, **69**, **70**
    dwellings by location **12**
    dwellings by tenure **7**
    high-rise flats 66, **66**, **74**
    housing repair costs **11**
    low-rise flats 66, **72**, **73**
    proportion of sash windows **21**
    proportion of slate roofs **20**
    proportions by wall structure **18**
    semi-detached houses **65**, **68**, **69**
    social housing 65, 66, **66**, **67**, **68**, **69**, **71**, **72**, **73**
    terraced houses **67**
    walls 65
    windows 65, **66**
eaves *133*
Edwardian period (1900–1918) **4**, 45–46, **45**
    Arts and Crafts **49**
    bungalows **53**
    construction types **8**
    converted flats **50**, **53**
    decorative features 45, **45**
    detached houses **52**, **53**
    dwellings by location **12**
    dwellings by tenure **7**
    Garden City-style homes 15, **15**, 36, **48**, **51**, **52**
    housing repair costs **11**

    low-rise flats **53**, **54**
    proportion of sash windows **21**
    proportion of slate roofs **20**
    proportions by wall structure **18**
    roofs 45, **45**, **49**, **50**
    semi-detached houses 45, **45**, **46**, **49**, **50**, **51**, **52**
    social housing **48**, **54**
    street names **16**, 19
    terraced houses **9**, **16**, **47**, **48**, **49**, **50**, **51**
    Tyneside flats **53**
    walls 46, **47**, **50**, **51**, **52**
    windows 45, **45**, **47**, **48**, **50**, **51**, **52**
'Edwardian'-style new homes **103**
energy efficiency
    Code for Sustainable Homes 95, 96
    interwar period (1919–1944) 56
    low-energy houses **91**, 95, **96**, **100**
    modern period (1981–2002) 86
    requirements 75, 76
    Victorian period (1850–1899) 35
energy performance certificates 95
English Heritage 15
entrance halls 45
Estates Action schemes 85, *134*

F

facades *118–119*, *134*
    early post-war period (1945–1964) 65
    Edwardian period (1900–1918) 45
    historic period (pre-1850) 25
    later post-war period (1965–1980) 75
    modern period (1981–2002) 86
    overcladding 9, **68**
    stucco 25, **30**, **34**, *119*, *135*
    tile hanging 65, 75, **77**, **79**, **117**, *119*
    timber boarding 65, **75**, **78**, **79**, **87**, **117**, *119*
    Victorian period (1850–1899) 35
    see also rendering
factory-built homes *see* system-built homes
fascias *133*
finials 45, *134*

fired clay building components 35, 45
flashings *114*, *129*
flat roofs **64**, 66, **78**, *110*, **111**, *113*
flats *see* converted flats; high-rise flats; low-rise flats
Flemish bonding 15, 19, **19**, *128*
flint 8, **38**, **117**, *118*
floor tiles 45
floorplans 15, **17**
floors
    insulation 86
    suspended timber 36, 46
foundations 35
four-in-a-block cluster homes **88**
four-in-a-block maisonettes **63**

G

gables and gable-ended roofs 66, *109*, *127*
    early post-war period (1945–1964) 66
    Edwardian period (1900–1918) 45, **50**
    half-timbered **50**, **63**
    interwar period (1919–1944) 56, **57**, **59**, **63**
    Victorian period (1850–1899) 36, **41**
Garden Cities Movement 15, **15**, 36, **48**, **51**, **52**
gardens 14, **14**
gated developments **89**, 95, **96**
geological map **25**
Georgian period (1760–1800) 15, **24**, **26**, **28**, **29**, **30**, **32**, **34**, *128*
'Georgian'-style houses **80**, **81**, **82**, **91**
glazed tiles 45
glazing *see* double glazing; triple glazing; windows
Goldfinger, Erno **66**
Gothic features 35, **40**
Gothic revival **34**
Great Fire of London 25
greenfield sites **14**, 75, 76, 85, **101**, **103**, **104**
gutters *114*, *129*

H

half-timbered effects **50**, **51**, **52**
heating systems 19
high-rise flats 2, **4–5**

high-rise flats (*continued*)
    early post-war period (1945–1964) 66, **66**, **74**
    later post-war period (1965–1980) 75, 76, **84**
    new homes (2003–2010) 96, **105**
hipped roofs 25, **55**, 56, 66, *109*, **111**, *134*
historic period (pre-1850) 3, **4**, 23–25, **23**, **24**
    construction types 8, **8**
    detached houses **26**, **30**, **31**, **32**
    dwellings by location 11, **12**
    dwellings by tenure **7**
    housing repair costs **11**
    proportion of sash windows **21**
    proportion of slate roofs **20**
    proportions by wall structure **18**
    semi-detached houses **23**, **32**
    terraced houses **26**, **27**, **28**, **29**
    townhouses **24**, **26**, **29**, **30**, **32**, **33**, **34**
    windows 25
history, local 15
'homes fit for heroes' 55
housing associations 7, 76
    modern period (1981–2002) 85
    new homes (2003–2010) 95
    numbers of dwellings **7**
Housing Defects Act (1984) 9
housing types 2–3, **4–5**

I

in situ concrete construction **10**, 55, **74**, *116*, *134*
    numbers of dwellings **8**
infill developments
    historic period (pre-1850) **27**
    new homes (2003–2010) 95–96, **99**, **103**, **104**
    small flats **92**, **93**
    terraced houses **16**
    urban blocks of flats **94**
insulation 8, 75, 86
integral garages **77**, **78**, **80**, **89**, **90**, **101**
internal services and fittings **18**, 19, 108, *130*
    historic period (pre-1850) 25
    interwar period (1919–1944) 56
    later post-war period (1965–1980) 75

interwar period (1919–1944) **4**, 55–56, **55**
    art deco-style housing 56, **56**, **61**, **62**, **64**
    bungalows **63**
    construction types **8**
    decorative features 56
    detached houses **60**, **61**, **62**
    dwellings by location **12**
    dwellings by tenure **7**
    housing repair costs **11**
    low-rise flats 56, **64**
    proportion of sash windows **21**
    proportion of slate roofs **20**
    proportions by wall structure **18**
    roofs **16**, 56, **57**, **58**, **59**, **61**, **63**
    semi-detached houses 14, **14**, **16**, **17**, 55, **55**, **56**, **58**, **59**, **60**, **61**
    social housing 55, 56, **57**, **58**, **60**, **61**, **64**
    street patterns and plot sizes 14, **14**
    terraced houses 55, **57**
    walls 56, **61**
    windows 56, **59**, **63**

J

jambs *130*
jettied storeys **27**, **28**, *127*

L

later post-war period (1965–1980) **5**, 75–76, **75**
    bungalows **82**, **83**
    construction types **8**, 9
    detached houses **76**, **78**, **80**, **81**, **82**
    dwellings by location **12**
    dwellings by tenure **7**
    high-rise flats 75, 76, **84**
    housing repair costs **11**
    low-rise flats 75, 76, **76**, **83**, **84**
    new towns 15
    proportion of sash windows **21**
    proportion of slate roofs **20**
    proportions by wall structure **18**
    semi-detached houses **75**, **78**, **79**
    social housing 75, 76, **76**, **77**, **83**, **84**
    terraced houses **77**, **78**, **80**
    walls 76

later post-war period (1965–1980) (*continued*)
    windows 75, **75**
'lattice' windows *131*
layouts 15, **17**
leaded lights
    casement windows **120**, *122*, *127*
    in doors *125*
Letchworth Garden City 15, **15**
Limehouse Basin, London **94**
lintels *116*, *131*, *134*
listed buildings 15, 23
local authorities
    local information 15
    see also council housing
local history 15
location 11, **12**
low-energy houses **91**, 95, **96**, **100**
low-rise flats 2, **4–5**, *135*
    built to look like houses **71**, **88**
    deck-access blocks 7, **7**, **64**, **74**, *134*
    early post-war period (1945–1964) 66, **72**, **73**
    Edwardian period (1900–1918) **53**, **54**
    in-fill developments **92**, **93**, **94**
    interwar period (1919–1944) 56, **64**
    later post-war period (1965–1980) 75, 76, **76**, **83**, **84**
    mansion blocks **44**, **64**, **74**
    modern period (1981–2002) **93**, **94**
    new homes (2003–2010) **99**, **100**, **101**, **103**, **104**, **105**
    over shops **33**, **44**, **54**, **72**, **73**
    Victorian period (1850–1899) 36, **36**, **43**, **44**

M

maisonettes **63**, **72**, **74**
mansard roofs **111**, *134*
mansion blocks **44**, **64**, **74**
mansions
    historic period (pre-1850) **33**, **34**
    Victorian period (1850–1899) **42**
maps, large-scale 15
market towns 15
materials 8–9
    vernacular 25, **26**, *135*
medieval buildings 15, 23
medieval features 35
metal-frame construction **10**, *116*

metal-frame construction (*continued*)
    interwar period (1919–1944) 55, 56
    numbers of dwellings **8**
    proportion by age **18**
metal roofs *109*, **111**, *113*
Methodist hall conversion **41**
mews-style houses **33**, **89**
mixed-use developments **96**, **105**
modern period (1981–2002) **5**, 85–86, **85**
    bungalows **92**
    construction types **8**
    detached houses **90**, **91**
    dwellings by location 11, **12**
    dwellings by tenure **7**
    housing repair costs **11**
    low-rise flats **93**, **94**
    numbers of dwellings built 85
    proportions by wall structure **18**
    semi-detached houses **85**, **88**, **89**
    social housing 85, **87**
    terraced houses **86**, **87**, **88**, **89**
    townhouses **86**, **89**, **90**, **92**, **93**

N

new homes (2003–2010) **5**, 95–96, **95**
    bungalows **101**
    construction types **8**, **10**
    detached houses **101**, **102**, **103**
    dwellings by location **12**
    dwellings by tenure **7**
    high-rise flats 96, **105**
    housing repair costs **11**
    low-energy houses 95, **96**, **100**
    low-rise flats **99**, **100**, **101**, **103**, **104**, **105**
    numbers of dwellings built 95
    proportions by wall structure **18**
    semi-detached houses **95**, **98**, **99**, **100**, **101**
    social housing 95, 96, **98**
    terraced houses **16**, **98**, **99**, **101**, **102**
    townhouses **98**, **100**, **101**
new towns 15
non-traditional construction *see* system-built homes

O

'Odeon'-style houses **60**
Old English bonding 15, 19, **19**
oriel windows **70**, 86, *134*
ornamentation *see* decorative features
overcladding 9, **68**

P

pantiles *133*
parapet roofs 25, **38**, **40**, *109*
party parapets **35**, *115*, *129*
'pavilion'-style bungalows **53**
Peabody Trust 36, **36**
pebble-dash rendering **55**, **57**, *117*, *119*
pediments *128*
picture windows 65
pilasters *134*
plaques 15, **16**
plastics 75
    see also PVC-U
plot sizes 7, 14, **14**
    new homes (2003–2010) **7**, 96, **101**, **102**, **103**
plumbing **57**, *114*
post-war period *see* early post-war period (1945–1964); later post-war period (1965–1980)
precast concrete construction 55, *116*, **117**, *134*
    numbers of dwellings **8**
    see also system-built homes
prefabricated homes 65
privately rented homes 7, **7**, 55
purpose-built flats *see* high-rise flats; low-rise flats; Tyneside flats
PVC-U
    doors **123**, *124*
    windows 19, 75, **120**, *121*

Q

Queen Anne revival style 35
quoins *128*

R

re-modelled houses **80**, **81**
reconstituted stone *115*
Regency period (1800–1830) 15, **15**, **24**, **28**, **29**, **33**
rendering **117**, *119*

rendering (*continued*)
    early post-war period (1945–1964) 65
    Edwardian period (1900–1918) 45, **47**, **50**, **52**
    historic period (pre-1850) **28**, 29
    interwar period (1919–1944) 55, **55**
    later post-war period (1965–1980) **78**
    pebble-dash **55**, **57**, **117**, *119*
    stucco 25, **30**, **34**, *119*, *135*
rent control 55
renting
    private 7, **7**, 55
    see also social housing
replacement windows **57**, **58**, **70**, **80**
ridge tiles 36, 45
    crenelated **111**, *134*
Right to Buy (RTB) scheme 85, *135*
roof lights *110*, *129*
roofs 19, *109–110*, **111**, *129*
    cat-slide **52**
    coverings **111**, *112–113*
    early post-war period (1945–1964) 66
    Edwardian period (1900–1918) 45, **45**, **49**, **50**
    flat **64**, 66, **78**, *110*, **111**, *113*
    hipped 25, **55**, 56, 66, *109*, **111**, *134*
    historic period (pre-1850) 25
    interwar period (1919–1944) **16**, 56, **57**, **58**, **59**, **61**, **63**
    later post-war period (1965–1980) 75
    mansard **111**, *134*
    metal *109*, **111**, *113*
    parapet 25, **38**, **40**, *109*
    thatched 19, **26**, **27**, **28**, **32**, **111**, *112*
    Victorian period (1850–1899) 36
    see also gables and gable-ended roofs; slate roofs; tiles, roof
room sizes
    Edwardian period (1900–1918) 46
    interwar period (1919–1944) 56
    modern period (1981–2002) 85
rough rendering 45, **47**, **52**
rural locations **12**
rusticated *135*

S

St Katharine Dock, London **15**
SAP *see* Standard Assessment Procedure (SAP)
sarking *113*
sash windows 19, **21**, **120**, *121*, *131*
    double-glazed **16**, **21**
    Edwardian period (1900–1918) 45, **45**, **48**, **51**, **52**
    interwar period (1919–1944) **16**
    proportion by age **21**
    Victorian period (1850–1899) **35**, 36
Saxon/Viking period (450–1066) **24**
segmental arch 35, **35**, **95**
semi-detached houses 2, **4–5**
    asymmetric pairs **79**
    'cottage-style' **51**
    early post-war period (1945–1964) **65**, **68**, **69**
    Edwardian period (1900–1918) 45, **45**, **46**, **49**, **50**, **51**, **52**
    'halls adjoining' style **60**
    historic period (pre-1850) **23**, 32
    interwar period (1919–1944) 14, **14**, **16**, **17**, 55, **55**, **56**, **58**, **59**, **60**, **61**
    later post-war period (1965–1980) **75**, **78**, **79**
    layouts **17**
    modern period (1981–2002) **85**, **88**, **89**
    new homes (2003–2010) **95**, **98**, **99**, **100**, **101**
    'Odeon'-style **60**
    street patterns and plot sizes 14, **14**
    Victorian period (1850–1899) **35**, **39**, **40**, **41**, **42**
    Wimpey-style **78**, **79**
servants' quarters 25, 36, 46
Shard, London **96**, **97**
shops, flats over **33**, **43**, **44**, **54**, **72**, **73**
sills *131*
single-glazed windows *121*, *131*, *133*
single-lap tiles *133*
skirtings 56, 75, *130*
slate cottages **39**
slate roofs **20**, 66, **111**, *112*, *129*
    artificial **16**, **20**, *102*, **111**, *112*
    interwar period (1919–1944) **16**, 56, **58**, **59**, **61**
    proportion by age **20**

social housing 7, **7**, 10
    Decent Homes schemes 85
    early post-war period (1945–1964) 65, 66, **66**, **67**, **68**, **69**, **71**, **72**, **73**
    Edwardian period (1900–1918) **48**, **54**
    Estates Action schemes 85
    interwar period (1919–1944) 55, 56, **57**, **58**, **60**, **61**, **64**
    later post-war period (1965–1980) 75, 76, **76**, **77**, **83**, **84**
    modern period (1981–2002) 85, **87**
    new homes (2003–2010) 95, **96**, **98**
    numbers of dwellings **7**
    Right to Buy (RTB) scheme 85
    Victorian period (1850–1899) 36, **36**, **43**
soffits *114*, *133*
solid masonry wall construction 8, 15, *115*
    Edwardian period (1900–1918) **50**, **51**, **52**
    numbers of dwellings **8**
    proportion by age **18**
    Victorian period (1850–1899) 36
spa towns 15
stained glass 45, **120**, *122*, *125*
Standard Assessment Procedure (SAP) 86, 95
steel-frame construction *see* metal-frame construction
stone construction 8, *115*, **117**
    Cotswold 8, **8**, **26**, **40**
    vernacular housing 25, **26**
    Victorian period (1850–1899) **37**, **38**, **40**, **41**
stonework 35, **35**
street names 19
    Edwardian period (1900–1918) **16**, 19
    Victorian period (1850–1899) 36
street patterns 14, **14**
stretcher bonding 15, **16**, 19, **19**, *132*
stucco 25, **30**, **34**, *119*, *135*
suburban locations **12**, 55, 56, 75
suspended timber floors 36, 46
sustainability standards 95, 96
    see also energy efficiency
system-built homes 8–9, 55, 65, 76, **76**
    Airey system 9, **9**
    BISF system 9, **10**
    Cornish system 9, **9**, **14**

Index

system-built homes (*continued*)
  numbers of dwellings 8
  'Wates' concrete panel system 68
system-built roofs *109*

T

tenure types 7, **7**
terraced houses 2, **4–5**
  back additions 15, **17**
  early post-war period (1945–1964) **67**
  Edwardian period (1900–1918) 9, **16**, 47, **48**, 49, **50**, **51**
  historic period (pre-1850) **26**, 27, **28**, **29**
  interwar period (1919–1944) 55, **57**
  later post-war period (1965–1980) **77**, **78**, **80**
  layouts 15, **17**
  modern period (1981–2002) **86**, **87**, **88**, **89**
  new homes (2003–2010) **16**, **98**, **99**, **101**, **102**
  staggered **51**, **77**, **80**
  street patterns and plot sizes 14, **14**
  Victorian period (1850–1899) 14, **14**, **17**, 35, 36, **36**, 37, **38**, **39**, **40**
terracotta building components 35, 45
thatched roofs 19, **26**, 27, **28**, 32, **111**, *112*
thermal insulation 8, 75, 86
tile hanging 65, 75, **77**, **79**, **117**, *119*
tiles, roof
  clay **45**, **111**, *112*, *127*
  concrete **47**, 66, 75, **85**, **111**, *112*, *133*
  ridge 36, 45, **111**, *134*
timber boarding 65, **75**, **78**, **79**, **87**, **117**, *119*
timber-frame construction *116*, *127*
  historic period (pre-1850) **8**, 25, **27**, **28**, **29**, **30**, **31**, **33**
  modern period (1981–2002) **85**, 86
  new homes (2003–2010) 9, **10**, **97**
  numbers of dwellings 8
  proportion by age 18
  wattle and daub infill **117**, *127*
timberwork, decorative 45, **45**
tower blocks *see* high-rise flats

townhouses
  historic period (pre-1850) 24, **26**, **29**, **30**, **32**, **33**, **34**
  modern period (1981–2002) **86**, **89**, **90**, **92**, **93**
  new homes (2003–2010) **98**, **100**, **101**
  Victorian period (1850–1899) **42**, **43**
triple glazing
  doors **123**, *124*
  windows **120**, *121*
tropical hardwoods 86, 96, **120**, *121*, *124*
Tudor/Elizabethan period (1485–1603) 24, **33**, *127*
'Tudorbethan' styling 36, 55, **56**, *135*
Tyneside flats 53
typology 2–3, **4–5**

U

urban locations 12

V

Venetian-style windows **26**, **32**
vernacular housing 25, **26**, *135*
Victorian Gothic 35, **40**
Victorian period (1850–1899) **4**, 35–36, **35**
  Arts and Crafts 36, **36**
  bungalows 42
  construction types 8, **8**
  converted flats 41, **42**, **43**, **44**
  decorative features 35, **37**
  detached houses **40**, **41**, **42**
  dwellings by location 12
  dwellings by tenure 7
  housing repair costs 11
  low-rise flats 36, **36**, **43**, **44**
  proportion of sash windows 21
  proportion of slate roofs 20
  proportions by wall structure 18
  semi-detached houses 35, **39**, **40**, **41**, **42**
  social housing 36, **36**, **43**
  street patterns and plot sizes 14, **14**
  terraced houses 14, **14**, **17**, 35, 36, **36**, 37, **38**, **39**, **40**
  townhouses **42**, **43**
  walls 36
  windows 35–36, **35**, **37**, **38**, **39**

village locations **12**

W

walk-up blocks of flats *135*
  early post-war period (1945–1964) 66, **73**
  interwar period (1919–1944) 56, **64**
  later post-war period (1965–1980) **83**
  modern period (1981–2002) **93**, **94**
wall tiling 45
walls 8, 15, *115–116*, **117**, *118–119*
  structure by age 18
  see also brick bonding; cavity wall construction; facades; metal-frame construction; solid masonry wall construction; stone construction; timber-frame construction
warehouse conversions **44**
'Wates' concrete panel system **68**
wattle and daub infill **117**, *127*
William IV houses **31**
Wimpey-style semi-detached houses **78**, **79**
windows 19, **120**, *121–122*, *131*
  dormer 36, 86, **95**, **120**, *133*
  double glazing **16**, **21**, 86, **120**, *121*
  early post-war period (1945–1964) 65, **66**
  Edwardian period (1900–1918) 45, **45**, **47**, **48**, **50**, **51**, **52**
  historic period (pre-1850) 25
  interwar period (1919–1944) 56, **59**, **63**
  later post-war period (1965–1980) 75, **75**
  metal-frame 65, **66**, **120**, *122*
  oriel **70**, 86, *134*
  PVC-U 19, 75, **120**, *121*
  replacement **57**, **58**, **70**, **80**
  stained glass 45, **120**, *122*, *125*
  Venetian-style **26**, **32**
  Victorian period (1850–1899) 35–36, **35**, **37**, **38**, **39**
  see also bay windows; casement windows; sash windows
wood shingles **111**, *112*
World War I 46, 55
World War II 65

Publications from IHS BRE Press

Fire performance of external thermal insulation for walls of multistorey buildings. 3rd edn. **BR 135**

External fire spread. 2nd edn. **BR 187**

Site layout planning for daylight and sunlight. 2nd edn. **BR 209**

Fire safety and security in retail premises. **BR 508**

Automatic fire detection and alarm systems. **BR 510**

Handbook for the structural assessment of large panel system (LPS) dwelling blocks for accidental loading. **BR 511**

Airtightness in commercial and public buildings. 3rd edn. **FB 35**

Biomass energy. **FB 36**

Environmental impact of insulation. **FB 37**

Environmental impact of vertical cladding. **FB 38**

Environmental impact of floor finishes: incorporating The Green Guide ratings for floor finishes. **FB 39**

LED lighting. **FB 40**

Radon in the workplace. 2nd edn. **FB 41**

U-value conventions in practice. **FB 42**

Lessons learned from community-based microgeneration projects: the impact of renewable energy capital grant schemes. **FB 43**

Energy management in the built environment: a review of best practice. **FB 44**

The cost of poor housing in Northern Ireland. **FB 45**

Ninety years of housing, 1921–2011. **FB 46**

BREEAM and the Code for Sustainable Homes on the London 2012 Olympic Park. **FB 47**

Saving money, resources and carbon through SMARTWaste. **FB 48**

Concrete usage in the London 2012 Olympic Park and the Olympic and Paralympic Village and its embodied carbon content. **FB 49**

A guide to the use of urban timber. **FB 50**

Low flow water fittings: will people accept them? **FB 51**

Evacuating vulnerable and dependent people from buildings in an emergency. **FB 52**

Refurbishing stairs in dwellings to reduce the risk of falls and injuries. **FB 53**

Dealing with difficult demolition wastes: a guide. **FB 54**

Security glazing: is it all that it's cracked up to be? **FB 55**

The essential guide to retail lighting. **FB 56**

Environmental impact of metals. **FB 57**

Environmental impact of brick, stone and concrete. **FB 58**

Design of low-temperature domestic heating systems. **FB 59**

Performance of photovoltaic systems on non-domestic buildings. **FB 60**

Reducing thermal bridging at junctions when designing and installing solid wall insulation. **FB 61**

Housing in the UK. **FB 62**

Delivering sustainable buildings. **FB 63**

Quantifying the health benefits of the Decent Homes programme. **FB 64**

The cost of poor housing in London. **FB 65**

Environmental impact of windows. **FB 66**

Environmental impact of biomaterials and biomass. **FB 67**

DC isolators for PV systems. **FB 68**

Computational fluid dynamics in building design. **FB 69**

Design of durable concrete structures. **FB 70**

A technical guide to district heating. **FB 72**

Available from www.brebookshop.com or call +44 (0) 1344 328038